C000258683

Toward HUMAN-LEVEL Artificial Intelligence

Representation and Computation of
Meaning in Natural Language

Philip C. Jackson, Jr.

Dover Publications, Inc.
Mineola, New York

Bibliographical Note

Toward Human-Level Artificial Intelligence: Representation and Computation of Meaning in Natural Language is a new work, first published by Dover Publications, Inc., in 2019.

International Standard Book Number

ISBN-13: 978-0-486-83300-2
ISBN-10: 0-486-83300-3

Manufactured in the United States by LSC Communications
83300301
www.doverpublications.com

2 4 6 8 10 9 7 5 3 1

2019

Dedication

To the memory of my parents, Philip and Wanda Jackson.

To my wife Christine.

Table of Contents

Contents

Figures

§ Notation and Overview of Changes

The § notation refers to chapters and sections in this book. For example, §2.1 refers to the first section in Chapter 2, which is labeled in the text and Table of Contents as 2.1. Its first subsection is §2.1.1.

This book combines text from a doctoral thesis with research papers based on the thesis, and elaborates some topics with further thoughts.

Relative to the thesis (Jackson, 2014) :

- The half-page Abstract has been replaced by a one-page Synopsis.

- New material was added in §1.5, §2.1.2.6, §2.1.2.9, §2.1.2.10, §2.1.2.11, §2.2.2, §2.2.4, §2.3.6, §2.3.7, §3.6.1, §3.6.7, §3.7.5, §4.2.5, §4.2.6, §5.3.

- §2.3.3.2.2 was moved into §4.2.6. §2.3.3.2.1 was moved up to §2.3.3.2.

- §2.3.3.6 is new. Previous material in §2.3.3.6 is now in §2.3.3.7.

- A new Chapter 8 has been added. Some material previously in Chapter 7 has been moved to §8.1 and §8.3. New material is added in §8.1, §8.2, §8.3.

- The previous Chapter 8 is now Chapter 9.

- New epigraphs have been used for some chapters.

- The infinity symbol is shown after each epigraph, to represent the potential scope of human-level artificial intelligence. Previously, each epigraph was followed by an icon for an open book.

- Quotations were removed where permissions did not cover a commercial book and possible translation to foreign languages.

- To improve readability, first-person pronouns are now used in several places, rather than references to "the author".

Synopsis

This book advocates an approach to achieve human-level artificial intelligence, based on a doctoral thesis (Jackson, 2014).

While a Turing Test may help recognize human-level AI if it is created, the test itself does not define intelligence or indicate how to design, implement, and achieve human-level AI.

The doctoral thesis proposes a design-inspection approach: to define human-level intelligence by identifying capabilities achieved by human intelligence and not yet achieved by any AI system, and to inspect the internal design and operation of any proposed system to see if it can in principle support these capabilities.

These capabilities will be referred to as *higher-level mentalities*. They include human-level natural language understanding, higher-level learning, metacognition, imagination, and artificial consciousness.

To implement the higher-level mentalities, the thesis proposes a novel research approach: Develop an AI system using a language of thought based on the unconstrained syntax of a natural language; Design the system as a collection of concepts that can create and modify concepts, expressed in the language of thought, to behave intelligently in an environment; Use methods from cognitive linguistics such as mental spaces and conceptual blends for multiple levels of mental representation and computation.

The thesis endeavors to address all the major theoretical issues and objections that might be raised against this approach, or against the possibility of achieving human-level AI in principle. No insurmountable objections are identified, and arguments refuting several objections are presented.

The thesis describes the design of a prototype demonstration system, and discusses processing within the system that illustrates the potential of the research approach to achieve human-level AI.

If it is possible to achieve human-level AI, then it is important to consider whether human-level AI *should* be achieved. So, this book discusses economic risks and benefits of AI, considers how to ensure that human-level AI and superintelligence will be beneficial to humanity, and identifies reasons why human-level AI may be necessary for humanity's survival and prosperity.

Preface

It is important to thank everyone who helped make the thesis possible, and who contributed to my research on artificial intelligence over the years, though time and space would make any list incomplete.

I am grateful to Professor Dr. Harry Bunt of Tilburg University and Professor Dr. Walter Daelemans of the University of Antwerp, for their encouragement and insightful, objective guidance of the thesis research and exposition. It was a privilege and a pleasure to work with them. I am also grateful to the other members of the thesis review committee for their insightful questions during the thesis defense in 2014: Dr. Filip A. I. Buekens, Professor Dr. H. Jaap ven den Herik, Professor Dr. Paul Mc Kevitt, Dr. Carl Vogel, and Dr. Paul A. Vogt.

Most doctoral dissertations are written fairly early in life, when memories are fresh of all who helped along the way, and "auld acquaintances" are able to read words of thanks. These words are written fairly late in life, regretfully too late for some to read.

I am grateful to all who have contributed directly or indirectly to my studies and research on artificial intelligence and computer science, in particular:

John McCarthy [1], Arthur Samuel, Patrick Suppes, C. Denson Hill, Sharon Sickel [2], Michael Cunningham, Ira Pohl, Edward Feigenbaum, Bertram Raphael, William McKeeman, David Huffman, Michael Tanner, Frank DeRemer, Ned Chapin, John Grafton, James Q. Miller, Bryan Bruns, David Adam, Noah Hart, Marvin Minsky, Donald Knuth, Nils Nilsson, Faye Duchin, Douglas Lenat, Robert Tuggle, Henrietta Mangrum, Warren Conrad, Edmund Deaton, Bernard Nadel, Thomas Kaczmarek, Carolyn Talcott, Richard Weyhrauch, Stuart Russell, Igor Aleksander, Helen Morton, Richard Hudson, Vyv Frederick Evans, Michael Brunnbauer, Jerry Hobbs, Laurence Horn, Brian C. Smith, Philip N. Johnson-Laird, Charles Fernyhough, Antonio

---

[1] McCarthy, Samuel, Suppes, and Hill were academic supporters of my Bachelor's program at Stanford – McCarthy was principal advisor.

[2] Sickel, Cunningham, and Pohl were academic supporters of my Master's program at UCSC – Sickel was principal advisor.

Chella, Robert Rolfe, Brian Haugh, K. Brent Venable, Jerald Kralik, Alexei Samsonovich, David J. Kelley, Peter Lindes, William G. Kennedy, Arthur Charlesworth, Joscha Bach, Patrick Langley, John Laird, Christian Lebiere, Paul Rosenbloom, John Sowa.

They contributed in different ways, such as teaching, questions, guidance, discussions, reviews of writings, permissions for quotations, collaboration, and/or correspondence. They contributed in varying degrees, from sponsorship to encouragement, lectures, comments, conversations, objective criticisms, disagreements, or warnings that I was overly ambitious. I profoundly appreciate all these contributions. To be clear, in thanking these people it is not claimed they would agree with everything I've written or anything in particular.

◊

It is appropriate to acknowledge the work of Noah Hart. In 1979, he asked me to review his senior thesis, on use of natural language syntax to support inference in an AI system. I advised the approach was interesting, and could be used in a system of self-extending concepts to support achieving human-level AI, which was the topic of my graduate research. Later, I forgot salient information such as his surname, the title of his paper, its specific arguments, syntax and examples, etc. It has now been over 39 years since I read his paper, which if memory serves was about 20 pages.

My research on the doctoral thesis initially investigated developing a mentalese based on conceptual graphs, to support natural language understanding and human-level AI. Eventually it was clear that was too difficult in the time available, because the semantics to be represented were at too high a level. So, I decided to explore use of natural language syntax, starting from first principles. Eventually it appeared this approach would be successful and, wishing to recognize Hart's work, I used resources on the Web to identify and contact him. He provided the title in the Bibliography, but said it was unpublished and he could not retrieve a copy. He recalled about his system[3]:

> "SIMON was written in Lisp and I had written a working prototype that was trained or 'taught'. There were hundreds of facts, or snippets of information initially loaded, and SIMON could respond to things it knew. It would also ask for

---

[3] Email from Noah Hart, December 2011.

more information for clarification, and ask questions as it tried to 'understand'."

To contrast, this doctoral thesis combines the idea of using natural language as a mentalese with other ideas from AI and cognitive science, such as the society of mind paradigm, mental spaces, and conceptual blends. The following pages discuss higher-level mentalities in human-level AI, including reflection and self-programming, higher-level reasoning and learning, imagination, and consciousness. The syntax for Tala presented here was developed without consulting Hart or referring to his paper. I recall he used a similar Lisp notation for English syntax, but do not recall it specifically.

◊

In general, my employment until retirement in 2010 was in software development and information technology. This was not theoretical research, though in some cases it involved working with other AI specialists on AI applications. I was fortunate to work with many of the best managers and engineers in industry, including Phil Applegate, Karen Barber, Doug Barnhart, Barbara Bartley, Ty Beltramo, Pete Berg, Dan Bertrand, Charles Bess, William Bone, Sam Brewster, Michelle Broadworth, Mark Bryant, Gregory Burnett, Tom Caiati, Pam Chappell, David Clark, David Coles, Bill Corpus, Justin Coven, Doug Crenshaw, Fred Cummins, Robert Diamond, Tom Finstein, Geoff Gerling, Dujuan Hair, Phil Hanses, Steve Harper, Kathy Jenkins, Chandra Kamalakantha, Kas Kasravi, Phil Klahr, Rita Lauer, Maureen Lawson, Kevin Livingston, David Loo, Steve Lundberg, Babak Makkinejad, Mark Maletz, Bill Malinak, Arvid Martin, Glenda Matson, Stephen Mayes, Stuart McAlpin, Eileen McGinnis, Frank McPherson, Doug Mutart, Bruce Pedersen, Tyakal Ramachandraprabhu, Fred Reichert, Paul Richards, Anne Riley, Saverio Rinaldi, Marie Risov, Patrick Robinson, Mike Robinson, Nancy Rupert, Bob Rupp, Bhargavi Sarma, Mike Sarokin, Rudy Schuet, Dan Scott, Ross Scroggs, Pradip Sengupta, Cheryl Sharpe, Scott Sharpe, Christopher Sherman, Michael K. Smith, Patrick Smith, Scott Spangler, Kevin Sudy, Saeid Tehrani, Zane Teslik, Kathy Tetreault, Lakshmi Vora, Rochelle Welsch, Robert White, Terry White, Richard Woodhead, Scott Woyak, Glenn Yoshimoto, and Ruth Zarger. I thank these individuals for leadership and collaboration. Again, any list would be incomplete and in thanking these people it is not claimed they would agree with everything I've written or anything in particular.

◊

It should be expressly noted that I alone am responsible for the content of this book. Naturally, I hope the reader will find that its value greatly outweighs its errors, and I apologize for any errors it contains.

I will always be grateful to my late parents, whose faith and encouragement made this effort possible. Heartfelt thanks also to other family and friends for encouragement over the years.

I'm especially grateful to my wife Christine, for her love, encouragement, and patience with this endeavor.

Philip C. Jackson, Jr.

Toward
HUMAN-LEVEL
Artificial
Intelligence

1. Introduction

> To unfold the secret laws and relations of those high
> faculties of thought by which all beyond the merely
> perceptive knowledge of the world and of ourselves is
> attained or matured, is an object which does not stand
> in need of commendation to a rational mind.

~ George Boole, *An Investigation of the Laws of Thought*, 1854

∞

1.1 Can Machines Have Human-Level Intelligence?

In 1950, Turing's paper on *Computing Machinery and Intelligence* challenged scientists to achieve human-level artificial intelligence, though the term *artificial intelligence* (AI) was not officially coined until 1955, in the Dartmouth summer research project proposal by McCarthy, Minsky, Rochester, and Shannon.

Turing suggested that scientists could say a computer thinks if it cannot be reliably distinguished from a human being in an "imitation game," which is now called a Turing Test. He suggested programming a computer to learn like a human child, calling such a system a "child machine," and noted that the learning process could change some of the child machine's operating rules. Understanding natural language would be important for human-level AI, since it would be required to educate a child machine and would be needed to play the imitation game.

McCarthy *et al.* proposed research toward computer systems that could achieve every feature of learning and intelligence. They proposed to investigate how computers could understand language, develop abstract concepts, perform human-level problem solving, and be self-improving. They planned to study neural networks, computational complexity, randomness and creativity, invention and discovery.

McCarthy proposed that his research in the Dartmouth summer project would focus on intelligence and language. He noted that every formal language yet developed omitted important features of English, such as the ability for speakers to refer to themselves and make statements about progress in problem-solving. He proposed to create a computer language that would have properties similar to English. The artificial language would allow a computer to solve problems by making conjectures and referring to itself. Concise English sentences would have equivalent, concise sentences in the formal language. McCarthy's envisioned artificial language would support statements about physical events and objects, and enable programming computers to learn how to perform tasks and play games.

Turing's 1950 paper concluded by suggesting two alternatives for developing machine intelligence. One alternative was to program a computer to play chess; the other was to create a child machine and teach it to understand and speak English.

The first approach, playing chess, was successfully undertaken by AI researchers, culminating in the 1997 victory of Deep Blue over the world chess champion Gary Kasparov. We[4] now know that this approach only scratches the surface of human-level intelligence. It is clear that understanding natural language is far more challenging: No computer yet understands natural language as well as an average five-year-old human child. No computer can yet replicate the ability to learn and understand language demonstrated by an average child.

Though Turing's paper and the Dartmouth proposal both stated the long-term research goal to achieve human-level AI, for several decades there were few direct efforts toward achieving this goal. Rather, there was research on foundational problems in a variety of areas such as problem-solving, theorem-proving, game-playing, machine learning, language processing, etc. This was perhaps all that could be expected, given the emerging state of scientific knowledge about these topics, and about intelligence in general, during these decades.

There have been many approaches, at least indirectly, toward the long-term goal. One broad stream of research to understanding intelligence has focused on logical, truth-conditional, model theoretic approaches to representation and processing, via predicate calculus, conceptual graphs, description logics, modal logics, type-logical semantics, and other frameworks.

A second stream of research has taken a bottom-up approach, studying how aspects of intelligence (including consciousness and

---

[4] In these pages, "we" often refers to the scientific community, or to people in general, e.g. "We now know X." It may also refer to the author plus the reader, e.g. "We next consider Y," or as a "royal we" to just the author, e.g. "We next present Z." Yet in no case does "we" refer to multiple authors; this thesis presents the doctoral research of just one author, P.C.J.

language understanding) may emerge from robotics, connectionist systems, etc., even without an initial, specific design for representations in such systems. A third, overlapping stream of research has focused on "artificial general intelligence," machine learning approaches toward achieving fully general artificial intelligence.

Parallel to AI research, researchers in cognitive linguistics have developed multiple descriptions for the nature of semantics and concept representation, including image schemas, semantic frames, idealized cognitive models, conceptual metaphor theory, radial categories, mental spaces, and conceptual blends. These researchers have studied the need for embodiment to support natural language understanding and have developed construction grammars to flexibly represent how natural language forms are related to meanings.

To summarize the current state of research, it has been clear for many years that the challenges to achieving human-level artificial intelligence are very great, and it has become clear that they are somewhat commensurate with the challenge of achieving fully general machine understanding of natural language. Progress has been much slower than Turing expected in 1950. He predicted that in fifty years people would commonly talk about machines thinking, and that this would be an educated opinion.

While people do informally speak of machines thinking, it is widely understood that computers do not yet really think or learn with the generality and flexibility of humans. While an average person might confuse a computer with a human in a typewritten Turing Test lasting only five minutes, there is no doubt that within five to ten minutes of dialog using speech recognition and generation (successes of AI research), it would be clear that a computer does not have human-level intelligence.

Progress on AI has also been much slower than McCarthy expected. In 2006 he gave a lecture in which he said he had hoped in 1955 that human-level AI would be achieved before many members of his audience were born.

Indeed, while many scientists continue to believe human-level AI will be achieved, some scientists and philosophers have for many years argued that the challenge is too great, that human-level AI is impossible in principle, or for practical reasons. Some of these arguments relate directly to elements of the approach of this thesis. Both the general and specific objections and theoretical issues will be discussed in detail, in

Chapter 4.

In sum, the question remains unanswered:

How could a system be designed to achieve human-level artificial intelligence?

The purpose of this thesis is to help answer this question, by describing a novel research approach to design of systems for human-level AI. This thesis will present hypotheses to address this question and present evidence and arguments to support the hypotheses.

1.2 Thesis Approach

Since the challenges are great, and progress has been much slower than early researchers such as Turing and McCarthy expected, there are good reasons to reconsider the approaches that have been tried and to consider whether another, somewhat different approach may be more viable. In doing so, there are good reasons to reconsider Turing's and McCarthy's original suggestions.

To begin, this thesis will reconsider Turing's suggestion of the imitation test for recognizing intelligence. While a Turing Test can facilitate recognizing human-level AI if it is created, it does not serve as a good definition of the goal we are trying to achieve, for three reasons. First, as a behaviorist test it does not ensure that the system being tested actually performs internal processing we would call intelligent. Second, the Turing Test is subjective: A behavior one observer calls intelligent may not be called intelligent by another observer, or even by the same observer at a different time. Third, it conflates human-level intelligence with human-identical intelligence. Rather than create human-identical AI, we may wish to create human-like, human-level AI. These issues are further discussed in §2.1.1 and §2.1.2.

This thesis will propose a different approach [5] that involves inspecting the internal design and operation of any proposed system to see if it can in principle support human-level intelligence. This approach defines human-level intelligence by identifying and describing certain capabilities not yet achieved by any AI system, in particular capabilities this thesis will call *higher-level mentalities,* which include natural language understanding, higher-level forms of learning and reasoning,

---

[5] A phrase describing this alternative as "augmenting" the Turing Test has been removed because the Turing Test focuses on AI indistinguishable from humans, rather than just human-like AI.

imagination, and consciousness.

Second, this thesis will reconsider Turing's suggestion of the child machine approach. Minsky (2006) gave a general discussion of this idea, also called the 'baby machine' approach. He said the idea has been unsuccessful because of problems related to knowledge representation: A baby machine needs to be able to develop new ways of representing knowledge, because it cannot learn what it cannot represent. This ability to develop new forms of representation needs to be very flexible and general.

It is not the case that people have been trying and failing to build baby machines for the past sixty years. Rather, as noted above, most AI research over the past sixty years has been on lower-level, foundational problems in a variety of areas such as problem-solving, theorem-proving, game-playing, machine learning, etc. Such research has made it clear that any attempts to build baby machines with the lower-level techniques would fail, because of the representational problems Minsky identified.

What we may draw from this is that the baby machine approach has not yet been adequately explored, and that more attention needs to be given to the architecture and design of a child or baby machine, and in particular to the representation of thought and knowledge. This provides motivation for Hypothesis I of this thesis (stated in §1.4 below), which describes a form of the baby machine approach. This thesis will discuss an architecture for systems to support this hypothesis and will make some limited progress in investigation of the baby machine approach. Chapters 3 and 4 will analyze theoretical topics related to this architecture and discuss how the approach of this thesis addresses the representational issues Minsky identified for baby machines.

Next, this thesis will reconsider approaches toward understanding natural language, because both Turing and McCarthy indicated the importance of natural language in relation to intelligence, and because it is clear that this remains a major unsolved problem for human-level AI. Indeed, this problem is related to Minsky's representational problems for baby machines, since the thoughts and knowledge that a human-level AI must be able to represent, and that a baby machine must be able to learn, include thoughts and knowledge that can be expressed in natural language.

Although McCarthy proposed in 1955 to develop a formal language with properties similar to English, his subsequent work did not exactly take this direction, though it appears in some respects he continued to pursue it as a goal. He designed a very flexible programming language, Lisp, for AI research, yet beginning in 1958 his papers concentrated on use of predicate calculus for representation and inference in AI systems, while discussing philosophical issues involving language and intelligence. In an unpublished 1992 paper, he proposed a programming language, to be called Elephant 2000, that would implement speech acts represented as sentences of logic. McCarthy (2008) wrote that the language of thought for an AI system should be based on logic, and gave objections to using natural language as a language of thought.

McCarthy was far from alone in such efforts: Almost all AI research on natural language understanding has attempted to translate natural language into a formal language such as predicate calculus, frame-based languages, conceptual graphs, etc., and then to perform reasoning and other forms of cognitive processing, such as learning, with expressions in the formal language. Some approaches have constrained and "controlled" natural language, so that it may more easily be translated into formal languages, database queries, etc.

Since progress has been very slow in developing natural language understanding systems by translation into formal languages, this thesis will investigate whether it may be possible and worthwhile to perform cognitive processing directly with unconstrained natural language, without translation into a conventional formal language. This approach corresponds to thesis Hypothesis II, also stated in §1.4 below. This thesis will develop a conceptual language designed to support cognitive processing of unconstrained natural language, in Chapters 3 and 5, and will discuss the theoretical ramifications of the approach. Chapter 4 will give a response to McCarthy's objections to use of natural language as a language of thought in an AI system, and to other theoretical objections to this approach.

Finally, in considering how to design a system that achieves the higher-level mentalities, this thesis will reconsider the relationship of natural language understanding to other higher-level mentalities and will consider the potential usefulness of ideas developed for understanding natural language, in support of higher-level mentalities. This approach corresponds to Hypothesis III of this thesis, also stated in §1.4 below. The thesis will make progress in investigation of this hypothesis, beginning in Chapter 3.

1.3 Terminology: Tala and TalaMind

To further discuss the approach of this thesis, it will be helpful to introduce some terminology to avoid cumbersome repetition of phrases such as "the approach of this thesis." (Other terms defined throughout the thesis are collected in the Glossary.)

The name *Tala*[6] refers to the conceptual language defined in Chapter 5, with the proviso that this is only the initial version of the Tala language, open to revision and extension in future work.[7] In general throughout this thesis, the word *concept* refers to linguistic concepts, i.e., concepts that can be represented as natural language expressions (cf. Evans & Green, 2006, p.158). The term *conceptual structure* will refer to an expression in the Tala conceptual language.

The name *TalaMind* refers to the theoretical approach of this thesis and its hypotheses, and to an architecture the thesis will discuss for design of systems according to the hypotheses, with the same proviso. TalaMind is also the name of the prototype system illustrating this approach.

1.4 TalaMind Hypotheses

The TalaMind approach is summarized by three hypotheses:

I. Intelligent systems can be designed as 'intelligence kernels', i.e. systems of concepts that can create and modify concepts to behave intelligently within an environment.

II. The concepts of an intelligence kernel may be expressed in an open, extensible conceptual language, providing a representation of natural language semantics based very largely on the syntax of a particular natural language such as English, which serves as a language of thought for the system.

---

[6] Trademarks for *Tala* and *TalaMind* have been created to support future development.

[7] The name *Tala* is taken from the Indian musical framework for cyclic rhythms, pronounced "Tah-luh," though I pronounce it to rhyme with "ballad" and "salad." The musical term *tala* is also spelled *taal* and *taala*, and coincidentally *taal* is Dutch for "language." Tala is also the name of the unit of currency in Samoa.

III.   Methods from cognitive linguistics may be used for multiple levels of mental representation and computation. These include constructions, mental spaces, conceptual blends, and other methods.

Previous research approaches have considered one or more aspects of these hypotheses, though it does not appear that all of them have been previously investigated as a combined hypothesis. For each hypothesis, the following pages will discuss its meaning and history relative to this thesis. The testability and falsifiability of the hypotheses are discussed in §1.6. Their relation to the Physical Symbol System Hypothesis is discussed in §1.4.4.

1.4.1   Intelligence Kernel Hypothesis

I.   Intelligent systems can be designed as 'intelligence kernels', i.e. systems of concepts that can create and modify concepts to behave intelligently within an environment.

This hypothesis is a description of a baby machine approach, stated in terms of conceptual systems, where concepts can include descriptions of behaviors, including behaviors for creating and modifying concepts. This hypothesis may be viewed as a variant of the Physical Symbol System Hypothesis (Newell & Simon, 1976), which is discussed in §1.4.4. It may also be viewed as a combination of the Knowledge Representation Hypothesis and the Reflection Hypothesis (Smith, 1982), which are discussed in §2.3.5, along with other related research.

Since I had written a book surveying the field of artificial intelligence published in 1974, upon entering graduate school in 1977 I decided to investigate how it might be possible to achieve "fully general artificial intelligence," AI at a level comparable to human intelligence. The resulting master's thesis (Jackson, 1979) formulated what is now Hypothesis I and discussed the idea of a self-extending intelligence kernel in which all concepts would be expressed in an extensible frame-based concept representation language. Hypotheses II and III of this thesis were not present in Jackson (1979).[8] It also did not envision the

---

[8] The wording in Jackson (1979) was "intelligent systems can be defined as systems of concepts for the development of concepts." It described an intelligence kernel as a system of initial concepts that could

TalaMind demonstration design and story simulations, which have been important for illustrating the TalaMind approach.

This thesis will investigate Hypothesis I by examining how executable concepts can be represented in natural language, and how an executable concept can create and modify an executable concept, within a story simulation. This will illustrate how behaviors can be discovered and improved, and how (as McCarthy sought in 1955) an AI system can refer to itself and formulate statements about its progress in solving a problem. There is much more work on intelligence kernels to be done in future research.

1.4.2 Natural Language Mentalese Hypothesis

> II. The concepts of an intelligence kernel may be expressed in an open, extensible conceptual language, providing a representation of natural language semantics based very largely on the syntax of a particular natural language such as English, which serves as a language of thought for the system.

This is a 'natural language of thought' hypothesis for human-level AI: an hypothesis that natural language syntax provides a good basis for a computer language of thought, and a good basis for representing natural language semantics. It disagrees with the view that English syntax is not important although semantics is important (§4.2.5), and posits instead that a natural language such as English is important because of how well its syntax can express semantics, and that the unconstrained syntax of a natural language may be used to support representation and processing in human-level AI.[9] The word *syntax* is

---

develop and extend its concepts to understand an environment, i.e. a self-extending system (viz. §2.3.5 re "seed AI"). The present wording embeds the definition of "intelligence kernel" within the hypothesis, and says "can be designed" rather than "can be defined," since a definition of something is different from a design to achieve it.

[9] To be clear, this thesis does not claim that people actually use English or other natural languages as internal languages of thought. Such claims are outside the scope of this thesis, which is focused only on how machines might emulate the capabilities of human intelligence. However, there is some evidence potentially supporting this hypothesis for human intelligence, briefly discussed in §2.2.4.

used in a very general sense, to refer to the structural patterns in a natural language that are used in communication.[10] This thesis will limit discussion of the hypothesis to the syntax of sentences, with topics such as morphology and phonology intended for future research.

The Tala conceptual language developed according to this hypothesis will have properties McCarthy initially proposed in 1955: It will support self-reference and conjecture, and its sentences will be as concise as English – since they will be isomorphic to English. As will be explained further beginning in §1.5, computer understanding of natural language semantics will require conceptual processing of the language of thought, relative to a conceptual framework and an environment. That is, understanding of semantics (and pragmatics in general) is a process that involves encyclopedic knowledge and at least virtual embodiment (an idea discussed in §2.2.3).

Fodor (1975) considered that a natural language like English might be used as a language of thought, extending a child's innate, preverbal language of thought. There is a long philosophical history to the idea of natural language as a language of thought, which this thesis does not attempt to trace. Even so, it appears there has been very little investigation of this idea within previous AI research. As noted in §1.2, research on natural language understanding has focused on translating natural language to and from formal languages. Russell and Norvig (2010) provide an introduction to the theory and technology of such approaches. While inference may occur during parsing and disambiguation, inference is performed within formal languages. Hobbs (2004) gives reasons in favor of first-order logic as a language of thought, discussed in §2.3.1. Wilks has advocated use of natural language for representing semantics, though his practical work has used non-natural language semantic representations. Section 2.2.1 discusses the 'language of thought' idea in greater detail.

Hart (1979, unpublished) discussed use of natural language syntax for inference in an AI system. Further information and acknowledgement are given in the Preface.

---

[10] The word *grammar* could be used instead, but has alternate senses that encompass linguistic meaning and knowledge of language (cf. Evans & Green, 2006, p.484).

1.4.3 Multiple Levels of Mentality Hypothesis

III. Methods from cognitive linguistics may be used for multiple levels of mental representation and computation. These include grammatical constructions, mental spaces, conceptual blends, and other methods.

This is an hypothesis that theoretical ideas developed for understanding natural language will be useful for achieving the higher-level mentalities of human-level intelligence, i.e. higher-level forms of learning and reasoning, imagination, and consciousness.

Hypothesis III was developed while working on this thesis. This hypothesis is equally as important as the first and second, and in some ways more important, since it identifies a direction toward achieving the higher-level mentalities of human-level intelligence, leveraging the first and second hypotheses. Of course, it does not preclude the use of other ideas from cognitive science to help achieve this goal.

This hypothesis is a result of pondering the multiple levels of mental representation and processing discussed by Minsky (2006), and considering how they could be represented and processed using a natural language mentalese. This led to the idea that the higher-level mentalities could be represented and processed within an intelligence kernel using a natural language mentalese with constructions, mental spaces, and conceptual blends. It does not appear that there is other, previous AI research exploring a hypothesis stated in these terms, where "multiple levels of mental representation and computation" includes the higher-level mentalities discussed in this thesis.

1.4.4 Relation to the Physical Symbol System Hypothesis

The TalaMind hypotheses are essentially consistent with Newell and Simon's (1976) Physical Symbol System Hypothesis (PSSH), which essentially hypothesizes that digital computers (or abstractly, physical symbol systems) can support human-level artificial intelligence. Briefly, Newell and Simon defined physical symbols as physical patterns, which can occur in expressions (symbol structures). A physical symbol system can contain a collection of expressions and have processes that operate on the expressions to produce new expressions. Expressions can designate processes to perform. The system can interpret expressions to perform the processes they designate. Newell and Simon noted that their definition of a physical symbol system essentially describes the symbolic processing abilities of digital computers.[11]

If the word "concept" is substituted for "expression," then a variant of PSSH is TalaMind Hypothesis I: "Intelligent systems can be designed as 'intelligence kernels', i.e. systems of concepts that can create and modify concepts to behave intelligently within an environment."

Newell and Simon stipulated that expressions can designate objects and processes. If expressions can also designate abstractions in general, then functionally there is not a difference between an expression and a conceptual structure, as the term is used in this thesis. The range of abstractions that can be designated in the Tala conceptual language is a topic discussed in Chapter 3.

In defining expressions as structures of symbols, PSSH implicitly suggests an intelligent system would have some internal language for its expressions. Newell and Simon discussed computer languages such as Lisp, and also mentioned natural language understanding as a problem for general intelligence. However, in discussing PSSH they did not hypothesize along the lines of TalaMind Hypotheses II or III, which are consistent with PSSH but more specific.

In presenting PSSH, Newell and Simon were not specific about the nature or definition of intelligence. They briefly said they were referring to the scope, abilities, behavior, speed, and complexity of human-level intelligence.

In §2.1.2 this thesis identifies specific features of human-level intelligence that need to be achieved in human-level AI.

1.5 TalaMind System Architecture

This thesis next introduces an architecture it will discuss for design of systems to achieve human-level AI, according to the TalaMind hypotheses. This is not claimed to be the only or best possible architecture for such systems. It is presented to provide a context for analysis and discussion of the hypotheses. Figure 1-1 on the next page shows elements of the TalaMind architecture. The term *Tala agent* will refer to a system with this architecture.

---

[11] Newell & Simon's definition of a physical symbol system appears to cover any programs that can be processed by a digital computer, including programs for neural networks (which they did not discuss). However, whether neural networks are covered by their definition of physical symbol systems is not central to the discussion of this thesis.

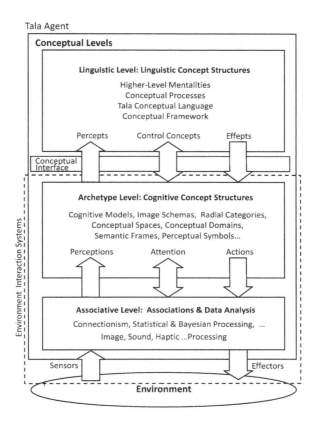

Figure 1-1 TalaMind System Architecture

In addition to the Tala conceptual language, the architecture contains two other principal elements at the linguistic level:

- *Conceptual Framework.* An information architecture for managing an extensible collection of concepts, expressed linguistically. A conceptual framework supports processing and retention of concepts ranging from immediate thoughts and percepts to long-term memory, including concepts representing definitions of words, knowledge about domains of discourse, memories of past events, context structures, etc.[12]

---

[12] In saying concepts are expressed linguistically, all methods are allowed, e.g. n-tuples of symbols, expressions in formal, logical languages,

- *Conceptual Processes*. An extensible system of processes that operate on concepts in the conceptual framework, to produce intelligent behaviors and new concepts.

Gärdenfors (1995) discussed three levels of inductive inference, which he called the linguistic, conceptual, and subconceptual levels. This thesis considers all three levels to be conceptual levels, due to its focus on linguistic concepts, and because an argument could be made that associative concepts exist. Hence the middle is called the *archetype level* to avoid describing it as the only conceptual level, and as a concise description that does not favor any particular cognitive concept representation. (It is not called the "cognitive level," since cognition also happens at the linguistic level, according to this thesis.) Section 2.2.2 further discusses the nature of concept representation at these levels. This thesis will discuss how the TalaMind architecture at the linguistic level could support higher-level mentalities in human-level AI.

In general, this thesis will not discuss the archetype and associative levels. Hence, throughout this thesis, discussions of "TalaMind architecture" refer to the linguistic level of the architecture, except where other levels are specified, or implied by context.

TalaMind is open to inclusion of other approaches toward human-level AI, for instance permitting predicate calculus, conceptual graphs, and other symbolisms in addition to the Tala language at the linguistic level, and permitting integration across architectural levels, e.g. potential use of neural nets at the linguistic and archetype levels. The TalaMind system architecture is actually a broad class of architectures, open to design choices at each level.

The TalaMind hypotheses do not require a generalized 'society of mind' architecture (§2.3.3.2) in which subagents communicate using the Tala conceptual language, but it is consistent with the hypotheses and natural to implement a society of mind at the linguistic level of the TalaMind architecture. This will be illustrated in Chapters 5 and 6.

This thesis does not discuss spatial reasoning and visualization,

---

or expressions in Tala. The framework should support representing expected future contexts, hypothetical or imaginary contexts, etc. These may be implemented using symbolic representations such as iconic mental models, which have been addressed in previous research (Johnson-Laird 1983 et seq.), discussed further in §2.3.6.

which may also occur in conceptual processing and are topics for future extensions of this approach.

From the perspective of the linguistic concept level, the lower two nonlinguistic levels of concept processing may be considered "environment interaction" systems. This interaction may be very complex, involving systems at the archetype level for recognizing objects and events in the environment, leveraging systems at the associative level, as well as sensors and effectors for direct interaction with the environment. While these environment interaction levels are very important, they are not central to this thesis, which will limit discussion of them and stipulate that concepts expressed in the Tala mentalese are the medium of communication in a *Conceptual Interface* between the linguistic level and the archetype level.

If environment interaction systems recognize a cat on a mat, they will be responsible for creating a mentalese sentence expressing this as a percept, received in the conceptual framework via the conceptual interface. If the conceptual processes decide to pet the cat on the mat, they will transmit a mentalese sentence describing this action via the conceptual interface to environment interaction systems responsible for interpreting the sentence and performing the action. This idea of a conceptual interface is introduced to simplify discussion in the thesis, and to simplify development of the thesis demonstration system: It enables creating a demonstration system in which Tala agents communicate directly with each other via the conceptual interface, abstracting out their environment interaction systems. As the TalaMind approach is developed in future research, the conceptual interface may become more complex; or, alternatively, it may disappear through integration of the linguistic and archetype levels. For instance, §§3.6.1 and 3.6.7.7 stipulate that concepts at the linguistic level can directly reference concepts at the archetype level.

In addition to action concepts ("effepts"), the linguistic level may send "control concepts" such as questions and expectations to the archetype level. For example, a question might ask the archetype level to find another concept similar to one it perceives, e.g. "What does the grain of wheat resemble?" and a percept might be returned, "The grain of wheat resembles a nut." Expectation concepts may influence what the archetype level perceives in information received from the associative level, and cause the archetype level to focus or redirect attention at the associative level. These are important topics, but they will be outside the focus of this thesis. Some discussion will be given related to them, in considering interactions between consciousness, unconsciousness, and understanding (§4.2.4).

This thesis relaxes PSSH requirements (2) and (3) stated by Newell and Simon (1976, p.116), by not requiring that all conceptual processes be describable in the Tala conceptual language, nor that all conceptual processes be alterable or created by other conceptual processes; it is allowed that some conceptual processes may result from lower-level symbolic or non-symbolic (associative) processing. Hence, TalaMind Hypothesis I may be considered a variant of PSSH.

1.6 Arguments and Evidence: Strategy and Criteria for Success

It should be stated at the outset that this thesis does not claim to actually achieve human-level AI, nor even an aspect of it; rather, it develops an approach that may eventually lead to human-level AI and describes a demonstration system to illustrate the potential of this approach.

Human-level artificial intelligence involves several topics, each so large that even one of them cannot be addressed comprehensively within the scope of a Ph.D. thesis. The higher-level mentalities are topics for a lifetime's research, and indeed, several lifetimes. Therefore, this thesis cannot claim to prove that a system developed according to its hypotheses will achieve human-level artificial intelligence. This thesis can only present a plausibility argument for its hypotheses.

To show plausibility, the thesis will:

- Address theoretical arguments against the possibility of achieving human-level AI by any approach.

- Describe an approach for designing a system to achieve human-level AI, according to the TalaMind hypotheses.

- Present theoretical arguments in favor of the proposed approach, and address theoretical arguments against the proposed approach.

- Present analysis and design discussions for the proposed approach.

- Present a functional prototype system that illustrates how the proposed approach could in principle support aspects of human-level AI if the approach were fully developed, though that would need to be a long-term research effort by multiple researchers.

After these elements of the plausibility argument are presented in Chapters 3 through 6, Chapter 7 will evaluate the extent to which they have supported the TalaMind hypotheses. Showing the plausibility of hypotheses will not be as clear-cut a result as proving a mathematical theorem, nor as quantitative as showing that a system can parse a natural language corpus with a higher degree of accuracy than other systems.

The general strategy of this thesis is to take a top-down approach to analysis, design, and illustration of how the three hypotheses can support the higher-level mentalities, since this allows addressing each topic, albeit partially. In discussing each higher-level mentality, the strategy is to focus on areas that largely have not been previously studied. Areas previously studied will be discussed if necessary to show it is plausible that they can be supported in future research following the approach of this thesis, but analyzing and demonstrating all areas previously studied would not be possible in a Ph.D. thesis. Some examples of areas previously studied are ontology, commonsense knowledge, encyclopedic knowledge, parsing natural language, uncertainty logic, reasoning with conflicting information, and case-based reasoning.

The success criterion for this thesis will simply be whether researchers in the field deem that the proposed approach is a worthwhile direction for future research to achieve human-level AI, based on the arguments and evidence presented in these pages.

The TalaMind approach is testable and falsifiable. There are theoretical objections that would falsify Hypothesis II and the Tala conceptual language. Some of these objections, such as Searle's Chinese Room Argument, would falsify the entire TalaMind approach and, indeed, all research on human-level AI. Objections of this kind are addressed in Chapter 4.

The Tala syntax defined in Chapter 5 could be shown to be inadequate by identifying expressions in English that it could not support in principle or with possible extensions. Tala's syntax has been designed to be very general and flexible, but there probably are several ways it can be improved.

Due to its scope, the TalaMind approach can only be falsified within a Ph.D. thesis by theoretical or practical objections, some of which are not specific to Tala. For example, the theoretical objections of Penrose against the possibility of achieving human-level AI would falsify the TalaMind approach, if one accepts them. Objections of this kind are also addressed in Chapter 4.

1.7 Overview of Chapters

Chapter 2 provides a review of previous research on human-level artificial intelligence and natural language understanding and proposes an alternative to the Turing Test, for defining and recognizing human-level AI. Chapter 3 will discuss the TalaMind architecture in more detail, to analyze theoretical questions and implications of the TalaMind hypotheses, and will discuss how a system developed according to the hypotheses could achieve human-level AI. Chapter 4 discusses theoretical issues and objections related to the hypotheses. Chapter 5 presents the design for a TalaMind prototype demonstration system. Chapter 6 describes processing within this system, which illustrates learning and discovery by reasoning analogically, causal and purposive reasoning, meta-reasoning, imagination via conceptual simulation, and internal dialog between subagents in a society of mind using a language of thought. The prototype also illustrates support for semantic disambiguation, natural language constructions, metaphors, semantic domains, and conceptual blends, in communication between Tala agents. Chapter 7 evaluates how well the preceding chapters support the hypotheses of this thesis. Chapter 8 discusses potential risks and benefits resulting from human-level artificial intelligence. Chapter 9 gives a summation of this thesis.

2. Subject Review: Human-Level AI and Natural Language

> Those who are enamoured of practice without science are like a pilot who goes into a ship without rudder or compass and never has any certainty where he is going. Practice should always be based upon a sound knowledge of theory, of which perspective is the guide and gateway, and without it nothing can be done well in any kind of painting.

> ~ Leonardo da Vinci, *Notebooks,* ca. 1510[13]

∞

2.1 Human-Level Artificial Intelligence

2.1.1 How to Define and Recognize Human-Level AI

As stated in §1.2, a Turing Test can facilitate recognizing human-level AI if it is created, but it does not serve as a good definition of the goal we are trying to achieve, for three reasons.

First, the Turing Test does not ensure that the system being tested actually performs internal processing we would call intelligent, if we knew what is happening inside the system. As a behaviorist test, it does not exclude systems that mimic external behavior to a sufficient degree that we might think they are as intelligent as humans, when they aren't.

For example, with modern technology we could envision creating a system that contained a database of human-machine dialogs in previous Turing Tests, with information about how well each machine response in each dialog was judged in resembling human intelligence. Initial responses in dialogs might be generated by using simple systems like Eliza (Weizenbaum, 1966), or by using keywords to retrieve information from Wikipedia, etc. The system might become more successful in passing Turing Tests over longer periods of time, simply by analyzing associations between previous responses and test results and giving responses that fared best in previous tests, whenever possible.

A system might also be designed to analyze all the publicly available

---

[13] From *Leonardo Da Vinci's Note-Books,* Arranged and rendered into English with Introductions by Edward McCurdy, M. A. (1906), New York: Charles Scribner's Sons.

information about a real human being, including the person's recorded interviews, speeches, and writings, and then imitate the person in a Turing Test, performing text processing without itself achieving human-level AI. Such a system might often be perceived as a real person in a Turing Test (cf. Peterson, 2013).

In 2011, a sophisticated information retrieval approach enabled the IBM Watson system to defeat human champions in the television quiz show *Jeopardy!* (Ferrucci *et al.*, 2010). A more limited technology using neural nets enables a handheld computer to successfully play "twenty questions" with a person (Burgener, 2006). Both of these are impressive, potentially useful examples of AI information retrieval, but they only demonstrate limited aspects of intelligence – they do not demonstrate true understanding of natural language, nor do they demonstrate other higher-level mentalities such as consciousness, higher-level reasoning and learning, etc.

The second reason the Turing Test is not satisfactory as a definition of human-level AI is that the test is subjective and presents a moving target: A behavior one observer calls intelligent may not be called intelligent by another observer, or even by the same observer at a different time. To say that intelligence is something subjectively recognized by intelligent observers in a Turing Test does not define where we are going, nor does it suggest valid ways to go there.

A third reason the Turing Test is not satisfactory is that it conflates human-level intelligence with human-identical intelligence, i.e. intelligence indistinguishable from humans. This is important, for instance, because in seeking to achieve human-level AI we need not seek to replicate erroneous human reasoning. An example is a common tendency of people to illogically chain negative defaults (statements of the form *Xs are typically not Ys*). Vogel (1996) examines psychological data regarding this tendency. Many other examples have been identified by Johnson-Laird (1983 *et seq.*), as will be discussed in §2.3.6.

Noting others had criticized the Turing Test, Nilsson (2005) discussed an alternative called the "employment test" to measure achievement of human-level AI by the percentage of jobs humans normally perform that can be performed by AI systems. Much earlier, Nilsson (1983, 1984) had warned about the potential for technological unemployment and discussed solutions to the problem that could be beneficial for humanity; this topic will be considered in §8.1.

While the employment test is an objective alternative to the Turing

Test, it too is a behaviorist test, with similar issues limiting its usefulness as a definition of human-level AI: Though most ordinary jobs require natural language understanding and commonsense reasoning, as well as domain-specific intelligence, arguably most do not require all the abilities of human-level intelligence to be discussed in the next section. It might not suffice to define the scope of the employment test as "all jobs" or "economically important jobs," because some abilities of human intelligence may be shown outside of employment, or may not be recognized as economically important.

Some AI researchers may respond to such definitional problems by, in effect, giving up and saying it is not possible to define human-level intelligence, even by external, behaviorist tests. Yet as discussed in §1.1, if we go back to the early papers of the field it is clear the original spirit of research was to understand every ability of human intelligence well enough to achieve it artificially. This suggests an intuition that it should be possible to have an internal, design-oriented explanation and definition of human-level intelligence.

The fact that we do not yet have an explanation or definition does not mean it is impossible or not worth seeking, or that human intelligence inherently must be defined by external, behaviorist tests. It may just mean we don't understand it well enough yet. The history of science is replete with things people were able to recognize, but for ages were unable to explain or define very well. This did not stop scientists from trying to understand. It should not stop us from trying to understand human intelligence well enough to define and explain it scientifically, and to achieve it artificially if possible.

Throughout the history of AI research, people have identified various behaviors only people could then perform, and called the behaviors "intelligent." Yet when it was explained how machines could perform the behaviors, a common reaction was to say they were not intelligent after all. A pessimistic view is that people will always be disappointed with any explanation of intelligent behavior. A more optimistic and objective response is to suppose that previously identified behaviors missed the mark in identifying essential qualities of human intelligence. Perhaps if we focus more clearly on abilities of human intelligence that remain to be explained, we will find abilities people still consider intelligent, even if we can explain how a computer could possess them. These may be internal, cognitive abilities, not just external behaviors. This will be endeavored, beginning in the next section.

Completeness is a very useful concept in this matter: People can always deny a system is intelligent, but one can always turn the table around and ask, "Can you show me something that in principle the system cannot do, which you or someone else can do?" Completeness arguments are a form of scientific falsifiability. If one can find something human intelligence can do that an AI system cannot, then a claim that the AI system is "human-intelligence complete" is falsified.

At present it is easy to find things existing AI systems cannot do. Perhaps someday that may not be the case. Perhaps someday a system will exist with such a complete design that no one will be able to find something that in principle it could not do, yet that humans can. Perhaps just by studying and testing its design and operation, reasonable people will arrive at the conclusion that it is human-intelligence complete, in the same way we say programming languages are Turing-complete because we cannot find any formal systems that exceed their grasp.

To summarize, an analysis of design and operation to say a system is human-intelligence complete would not be a behaviorist test. It would be an analysis that supports saying a system achieves human-level artificial intelligence, by showing its internal design and operation will support abilities we would say demonstrate human-level intelligence, even when we understand how these abilities are provided.

2.1.2 Unexplained Features of Human-Level Intelligence

Given the previous discussion, this section lists some of the unexplained characteristics of human-level intelligence, concentrating on essential attributes and abilities a computer would need to possess human-level artificial intelligence.

2.1.2.1 Generality

A key feature of human intelligence is that it is apparently unbounded and completely general. Human-level AI must have this same quality. In principle there should be no limits to the fields of knowledge the system could understand, at least so far as humans can determine.

Having said this, it is an unresolved question whether human intelligence is actually unbounded and completely general. Some discussion related to this is given in Chapter 4. Here it is just noted that while we may be optimistic that human intelligence is completely general, there are many limits to human understanding at present. For instance:

- Feynman at times suggested quantum mechanics may be inherently impossible for humans to understand, because experimental results defy commonsense causality. Yet at least people have been able to develop a mathematical theory for quantum mechanics, which has been repeatedly verified by experiments, to great precision.

- General relativity and quantum theory are not yet unified. Astronomers have evidence black holes exist, which implies existence of gravitational singularities.

- At present scientists are having great difficulty explaining multiple, independent observations that appear to prove 95% of the universe consists of matter and energy we have not yet been able to directly observe, causing galaxies and galaxy clusters to rotate faster than expected, and causing the expansion of the universe to accelerate (Gates, 2009).

- Beyond this, there are several other fundamental questions in physics one could list that remain open and unresolved. And there are many open, challenging questions in other areas of science, including the great question of precisely how our brains function to produce human intelligence.

There is no proof at this point that we cannot understand all the phenomena of nature. And as Chapter 4 will discuss, it is an unsettled question whether human-level artificial intelligence cannot also do so. Hopefully human-level AI will help us in the quest. Research on AI systems for discovery of scientific theories is presented in (Langley *et al.*, 1987) and in (Shrager and Langley, 1990).

2.1.2.2 Creativity and Originality

A key feature of human intelligence is the ability to create original concepts. Human-level AI must have this same quality. The test of originality should be whether the system can create (or discover, or accomplish) something for itself it was not taught directly – more strongly, in principle and ideally in actuality, can it create something no one has created before, to our knowledge? This is Boden's (2004) distinction of (personal, psychological) P-creativity vs. (historical) H-creativity.

2.1.2.3   Natural Language Understanding

A key feature of human intelligence is the ability to understand natural languages, such as English or Dutch. Understanding natural language is still largely an unexplained skill of human-level intelligence. Attempts to build systems that process natural language have made substantial progress in many areas of syntax processing, but they still founder on the problem of understanding natural language in a general way.

2.1.2.4   Effectiveness, Robustness, Efficiency

The system must be effective in solving problems and behave successfully within its environment. The system must be able to deal with conflicting and uncertain information. The system must be able to identify and correct logical errors. The system must be able to rapidly acquire human-level knowledge, and deal with intractable domains and large amounts of information, at least as well as people do.

These are very important requirements for eventually achieving human-level artificial intelligence, but they will only be discussed in this thesis relative to its primary focus, to show how higher-level mentalities can in principle be supported by the TalaMind approach. Hence, this thesis will be more concerned with effectiveness and robustness than with efficiency and scalability, e.g. because we will need to discuss how a system that reasons with a natural language mentalese can detect and resolve contradictions. Efficiency and scalability issues will be noted in discussing other topics, but work on them will be a major topic for future research.

2.1.2.5   Self-Development and Higher-Level Learning

A variation of the requirement for originality is a requirement for 'self-development'. People not only discover new things, they develop new skills they were not taught by others, new ways of thinking, etc. A human-level AI must have this same capability. More specifically, human-level intelligence includes the following higher-level forms of learning:

- o Learning by induction, abduction, analogy, causal and purposive reasoning.
    - Learning by induction of new linguistic concepts.

- Learning by creating explanations and testing predictions, using causal and purposive reasoning.
- Learning about new domains by developing analogies and metaphors with previously known domains.

o Learning by reflection and self-programming.

- Reasoning about thoughts and experience to develop new methods for thinking and acting.
- Reasoning about ways to improve methods for thinking and acting.

o Learning by invention of languages and representations.

We shall use the term *higher-level learning* to describe these collectively and distinguish them from lower-level forms of learning investigated in previous research on machine learning (viz. Russell & Norvig, 2010).

2.1.2.6 Metacognition and Multi-Level Reasoning

Metacognition[14] is "cognition about cognition," cognitive processes applied to cognitive processes. This does not say much, until we say what we mean by cognition. There are both broad and narrow usages for the term *cognition* in different branches of cognitive science and AI. Many authors distinguish cognition from perception and action.

However, Newell (1990, p.15) gave reasons why perception and motor skills should be included in "unified theories of cognition." If we wish to consider metacognition as broadly as possible, then it makes sense to start with a broad idea of cognition, including perception, reasoning, learning, and acting, as well as other cognitive abilities Newell identified, such as understanding natural language, imagination, and consciousness.

Since cognitive processes may in general be applied to other cognitive processes, we may consider several different forms of metacognition, for example:

Reasoning about reasoning.
Reasoning about learning.

---

[14] This section uses text I provided as input to a joint paper (Kralik *et al.*, 2018), which summarizes several other approaches to metacognition.

Learning how to learn.

...

Others have focused on different aspects of metacognition, such as "knowing about knowing" or "knowing about memory." Cognitive abilities could be considered in longer metacognitive combinations, e.g., "imagining how to learn about perception" – the combination could be instantiated to refer to a specific perception.

Such examples illustrate that natural language has syntax and semantics which can support describing different forms of metacognition. More importantly, a 'natural language of thought' could help an AI system perform metacognition by supporting inner speech (§2.2.4) and by enabling the expression of specific thoughts about other specific thoughts, specific thoughts about specific perceptions, etc.

While in principle one may argue that all forms of metacognition could be supported by the TalaMind approach, attention in this thesis is focused on inner speech and meta-reasoning (reasoning about reasoning). For concision, the term *multi-level reasoning* will be used to refer collectively to the reasoning capabilities of human-level intelligence, including meta-reasoning, analogical reasoning, causal and purposive reasoning, abduction, induction, and deduction. It remains a challenge to include multi-level reasoning in a unified framework for human-level artificial intelligence, integrated with other unexplained features of intelligence.

2.1.2.7 Imagination

Imagination allows us to conceive things we do not know how to accomplish, and to conceive what will happen in hypothetical situations. To imagine effectively, we must know what we do not know, and then consider ways to learn what we do not know or to accomplish what we do not know how to do. A human-level AI must demonstrate imagination.

2.1.2.8 Consciousness

To act intelligently, a system must have some degree of awareness and understanding of its own existence, its situation and relation to the world, and its perceptions, thoughts, and actions, both past and present, as well as potentials for the future. Without such awareness, a system is greatly handicapped in managing its interactions with the world, and in managing its thoughts. So, at least some aspects of consciousness are necessary for a system to demonstrate human-level intelligence.

In stating this requirement, this thesis goes beyond what has been a standard assumption of many AI researchers: Turing (1950) wrote that the question of whether machines can think could be answered without solving the mystery of consciousness. Russell and Norvig (2010) agreed with Turing that we can create intelligent programs without trying to make them be conscious.

On the other hand, both McCarthy (1995, 2002) and Minsky (2006) have discussed how AI systems might emulate aspects of consciousness. Section 2.3.4 will discuss research on "artificial consciousness" conducted by Aleksander *et al.* (1992 *et seq.*) and others.

The perspective here is that it is both necessary and possible for a system to demonstrate at least some aspects of consciousness, to achieve human-level artificial intelligence. This thesis accepts the objection of AI critics that a system that is not aware of what it is doing, and does not have some awareness of itself, cannot be considered to have human-level intelligence. Further, consciousness is intertwined with understanding of natural language, and understanding in general, as we shall see in §4.2.4's discussion of Searle's Chinese Room Argument.

2.1.2.9 Sociality, Emotions, Values

A human-level AI will need some level of social understanding to interact with humans. It will need some understanding of cultural conventions, etiquette, politeness, etc. It will need some understanding of emotions humans feel, and it may even have some emotions of its own, though we will need to be careful about this. One of the values of human-level artificial intelligence is likely to be its objectivity and freedom from being affected by some emotions. People would be very concerned about interacting with emotional robots if robots could lose control of their emotions and become emotionally unpredictable. We probably would not want an AI system performing an important function like air traffic control to be emotional. On the other hand, we might want a robot taking care of infants, children, or hospital patients to show compassion and affection (cf. McCarthy, 2004); we might want a robot defending a family from violent home invaders to emulate anger.

Within an AI system, emotions could help guide choices of goals, or prioritization of goals. Apart from whether and how emotions may be represented internally, a human-level AI would also need to understand how people express emotions in behaviors and linguistically, and how its behaviors and linguistic expressions may affect people and their emotions.

A human-level AI must have values that guide its efforts to understand and act within its environment, and with human beings. It must have some understanding of human values to interact successfully with us. Thus, a human-level AI will need an understanding of ethical values, ethical rules, and principles to interact with humans, and to support "beneficial AI" – AI that is beneficial to humanity and to life in general (Bringsjord, Arkoudas, & Bello, 2006; Tegmark, 2017). This topic has become increasingly important as people have considered the potential good and bad consequences AI might have for humanity.

Questions related to sociality, emotions, and values are even more difficult and at a higher level than the issues that are the primary focus of this thesis. Section 3.7.5 will give a few very preliminary remarks about this topic, within the TalaMind approach. Section §8.2 discusses future issues related to beneficial AI in more detail.

2.1.2.10 Visualization, Spatial-Temporal Reasoning

Very closely related to imagination (some might claim identical) is the ability people have to visualize situations in three-dimensional space and reason about how these situations might change, e.g. by visualizing motions of objects. This ability is important for understanding natural language expressions and metaphors, for imagination, and for discovery of theories and inventions. So, this ability is listed as a higher-level mentality of human-level intelligence, though arguably it is foundational for cognition in general. Visualization and spatial-temporal reasoning are topics for future research and development in the TalaMind approach.

2.1.2.11 Curiosity, Self-Programming, Theory of Mind

To support higher-level learning, an intelligent system must have another general trait, *curiosity*, which at the level of human intelligence may be described as the ability to ask relevant questions and understand relevant answers.

In English, questions involve the interrogatives *who, what, where, when, why,* and *how.* The last two in particular merit further discussion:

A *how* question asks for a description of a method, which can be a procedure or a process. To understand the answer, an intelligent system needs to be able to represent procedures and processes, think about such representations, and ideally perform the procedures or processes described by representations, if it has the necessary physical abilities and resources. It is natural for an intelligent system to represent procedures and processes at the linguistic level of its AI architecture. With such representations it is a relatively direct step to support self-programming within an AI system.

A *why* question asks for a description of either a cause or an intent. Understanding the answer requires that an intelligent system be able to support causal reasoning about physical events, and also be able to support reasoning about people's intentions for performing actions. Reasoning about intentions involves supporting 'Theory of Mind', the ability for an AI system to consider itself and other intelligent agents (including people) as having minds with beliefs, desires, different possible choices, etc.

2.1.2.12 Other Unexplained Features

In addition, there are other features of human-level intelligence one could imagine eventually wishing to address in artificial intelligence, yet which are even more difficult and remote from consideration at this time.

One such feature is "freedom of will." This is a difficult philosophical topic, with debate about its nature and whether humans truly possess it in a universe apparently predetermined by the laws of physics. It will be a topic outside the scope of this thesis.

Beyond emotions, values, and freedom of will, unexplained features include "virtues." There may be no reason in principle why we would not want an artificial intelligence to possess a virtue such as wisdom, kindness, or courage, if the situation merited this. Yet what little wisdom I possess indicates it would not be wise to discuss wisdom or other virtues in this thesis. It is challenging enough to discuss higher-level mentalities such as imagination and consciousness.

2.2 Natural Language

2.2.1 Does Thought Require Language?

This is an old and important question. For example, Wittgenstein (1953) wrote that St. Augustine [15] described the language learning

---

[15] Viz. Wittgenstein (1953, p.15e, remark #32), and Augustine's *Confessions*, Book I, Chapter VIII paragraph 13. Augustine also suggests

process as if a child has an innate language preceding and enabling the acquisition of a spoken, public language.

Wittgenstein's own thoughts about the relationship between thought and language may be difficult to discern, because he discusses the topic throughout his *Philosophical Investigations* in a series of Socratic questions to himself and the reader, often seeming to answer each question with another question. Mulhall (2007) notes that Wittgenstein is open to the idea that an individual may talk to himself but questions whether an individual can have a private language to express inner experiences that are inherently private, such as sensations of pain. It does not appear that Wittgenstein considered the possible role of a language of thought in a society of mind (§2.3.3.2), i.e. it appears he took the unity of self as axiomatic.

Fodor (1975 *et seq.*) argued in favor of a language of thought hypothesis, essentially in agreement with Augustine. This has been the subject of lengthy philosophical arguments pro and con, e.g. concerning issues such as whether an innate language is needed to learn an external language and the degree to which an innate language must contain all possible concepts, or constrains the concepts that can be learned and expressed. Fodor (2008) accepted the principle of semantic compositionality, an issue in earlier philosophical debates. Fodor (1975) considered that a natural language like English might be used as a language of thought, extending a child's innate, preverbal language of thought. He reasoned the innate language of thought must be as powerful as any language learnable by humans, though extensions such as English would enable concise expression of concepts not primitive in the innate language. He also described the innate language of thought as a meta-language, in which natural language extensions could be defined.

Fodor's writings do not yield the only possible language of thought theory. Schneider (2011) considered arguments for and against Fodor's theory and presented an alternative theory for a computational language of thought, which she developed to be compatible with cognitive science and neuroscience.

---

humans have an innate gestural natural language, which supports learning of spoken natural languages – an idea being explored in modern work, e.g. by Sloman (2008). See also Tomasello (2003) regarding the importance of gestures for acquiring natural language.

Sloman (1979 *et seq.*) contended that the primary role of language is the representation of information within an individual, and that its role in communication is an evolutionary side effect, i.e. human-level intelligence requires some innate, internal language for representation of thoughts, prior to learning and using natural language. (Viewed this way, the existence of an internal representation language can be seen as a corollary of the Physical Symbol System Hypothesis.) Sloman disagreed with Fodor about the necessary content of the innate language, arguing that in principle a system can learn new concepts (which may be represented by new words or symbols) that may not be definable in terms of previously known concepts, words, or symbols. Thus, he emphasized the extensibility of innate representation languages.

Nirenburg and Wilks (2001) give a dialog on questions about ontologies, representations, and languages. Wilks essentially argues that representation languages (RLs) are natural languages (NLs) in some respects. Nirenburg argues against this. Wilks suggests that the predicates of any semantic representation language will either inherently or eventually represent natural language words, and have the ambiguity of NL words. Nirenburg contends that predicates can be defined as distinct senses of NL words. This is consistent with Wilks' previous theoretical work and with the view of Wittgenstein (and some of his other followers, e.g. Masterman and Spärck Jones) that the meaning of words depends on how they are used.

Berwick and Chomsky (2016) give a perspective on the evolution and nature of a language of thought in humans and discuss how it might be related to an innate "universal grammar" (Chomsky, 1966). There are diverse theories for how children learn languages (e.g. Vygotsky, 2012; Piaget, 1926) and for the evolution of language in our species (e.g. Coulardeau & Eve, 2017).

It is tempting to say that if we restrict "language" to verbal or written, serial human natural languages such as English, Chinese, etc., then thought is possible without language: People can solve some kinds of problems using spatial reasoning and perception that are at least not easy to express in English. Children can display intelligence and thinking even if they haven't yet learned a language such as English. Pinker (1994) cites medical and psychological evidence showing that thought and intelligence are not identical to the ability to understand spoken, natural languages. Yet these considerations do not rule out the possibility that a child's mind may use an innate language of thought to support reasoning, before the child learns a spoken natural language.

Pinker also argues against the Sapir-Whorf hypothesis that language determines and limits our thinking abilities, providing a variety of arguments and evidence to refute a strict interpretation of Sapir-Whorf. On the other hand, Boroditsky and Prinz (2008) discuss evidence that statistical regularities in English, Russian, and other natural languages have an important role in thought, suggesting people who speak different languages may think in different ways. And Pinker (1994, p.72) concluded that people do have a language of thought, or *mentalese*, though he reasoned that it is different from a spoken, natural language.

There is an elegant argument that concepts must be expressed as sentences in a mental language (viz. Jackendoff, 1992): Since natural language sentences can describe an effectively unlimited number of concepts, and the brain is finite, concepts must be represented internally within the mind as structures within a combinatorial system, or language.[16] Jackendoff called these concepts "sentential concepts." He developed a theory of *conceptual semantics* to provide a linguistic description of concepts corresponding to the semantics of natural languages (Jackendoff, 1983 *et seq.*).

Pinker (2007, p.150) agrees human intelligence may rely on conceptual semantics as an internal language of thought distinct from spoken natural languages. Spoken natural languages may be seen as ways of "serializing" mentalese concepts for communication between people. The psychological experiments cited against the equivalence of language and thought may only show cases where the mechanisms for spoken language are impaired, while the mechanisms for mentalese continue to function, or vice versa.

The expressive capabilities of natural languages should be matched by expressive capabilities of mentalese, or else by Jackendoff's argument the mentalese could not be used to represent the concepts expressed in natural language. The ability to express arbitrarily large, recursively structured sentences is plausibly just as important in a mentalese as it is in English. The general-purpose ability to metaphorically weld concepts together across arbitrary, multiple domains is plausibly just as important in a mentalese as it is in English. Considering Jackendoff's argument, it is cognitively plausible that natural language

---

[16] Others gave similar arguments, e.g. Chomsky (1975), Fodor (1975).

representation and processing are in some form *core functionalities* of human-level intelligence, needed for representation of thoughts.

This is not to say mentalese would have the same limitations as spoken English, or any particular spoken natural language. In mentalese, sentences could have more complex, non-sequential, graphical structures not physically permitted in speech, and indeed this thesis will use hierarchical list structures for representing English syntax, to facilitate conceptual processing.

2.2.2   What Does Meaning Mean?

To address this question, this section briefly discusses Peirce and Wittgenstein's theories of understanding and meaning. Wilks *et al.* (1996a) survey the history of thoughts about meaning. Nirenburg and Raskin (2004) discuss the evolution of formal representations for semantics and ontologies.

Besides understanding natural language, Peirce also considered understanding of phenomena in general, e.g. developing and using explanations of how (by what cause) and why (for what purpose) something happens or is done. Peirce discussed language as a system of signs, where a 'sign' is something that can stand for (represent) something else.

Peirce described a general process by which signs are understood. He called an initial sign (thing to be understood) a *representamen*. It is typically something external in the environment. It may be a symbol printed on paper (such as a Chinese symbol for "lamp" 灯); or smoke perceived at a distance; or, to use Atkin's (2010) example, a molehill in one's lawn; or a natural language utterance (such as "a log is in the fireplace"); or anything else perceived in the environment.

The process of understanding the representamen leads the mind to conclude that it stands for (or represents, or suggests the existence of) something, called the *object*. The object of the Chinese symbol might be a real lamp, the object of the smoke might be a fire that produces it, the object suggested by the molehill could be a mole that created it, the object of the natural language utterance could be a log in a fireplace, etc.

From Peirce's perspective, the process of understanding a sign or representamen involves developing an explanation for the meaning or cause of the sign. Peirce used the term *abduction* to refer to reasoning that develops explanations: If one observes something surprising, B, then one considers what fact A might naturally cause or explain B, and one concludes it is reasonable to think A might be true (Peirce, CP 5.189).

So, understanding involves developing explanations for what is observed. This applies both to understanding natural language and to understanding in general for human intelligence (cf. Hobbs *et al.*, 1993; Bunt & Black, 2000).

According to Peirce, the mind does not go directly from the representamen to the object in developing an explanation for what is observed. The mind internally creates another sign, called the *interpretant*, which it uses to refer to the object. Within the mind, the interpretant stands for, or represents, the external object that is the represented by the first sign, the representamen (Peirce, CP 2.228).[17]

We do not have to know precisely how this internal sign is expressed in the brain to believe some pattern of physical information must exist in the brain constituting an internal sign, providing a link between the external representamen and the external object. Importantly, we do not have to believe there is just one kind of physical information pattern used to express all internal meanings – the brain could use a variety of different physical information media and patterns for expressing meanings.[18]

Though Wittgenstein (1922) presented a purely logical description of the relationship between language and reality in *Tractatus Logico-Philosophicus*, he later restated much of his philosophy about language in *Philosophical Investigations*. A central focus of *Investigations* was the idea that the meaning of words depends on how they are used, and that words in general do not have a single, precisely defined meaning. As an example, Wittgenstein considered the word "game" and showed it has many different, related meanings. What matters is that people are able to use the word successfully in communication about many different things. Wittgenstein introduced the concept of a "language game" as an activity in which words are given meanings according to the roles that

---

[17] Viz. Atkin's (2010) discussion of how Peirce's theory of signs evolved throughout his lifetime. Vogt (2000 *et seq.*) has used computer simulation of the Peircean triad in studies of symbol grounding and language learning; also see Vogel & Woods (2006).

[18] The original text of this paragraph used "represent" instead of "constitute" and "express," which are now used to clarify and avoid over-using "represent."

words perform in interactions between people.[19]

It does not appear there is any fundamental contradiction between Wittgenstein and Peirce. Rather, what Wittgenstein emphasized was that an external representamen may stand for many different external objects. From a Peircean perspective this implies that the representamen may have many different internal signs, or interpretants, corresponding to different external meanings in different situations. A Peircean understanding process needs to support disambiguation (via abductive inference) of different interpretants to understand what a usage of an external sign means in a particular context.

These considerations can be summarized by saying that just as a word like "game" can have many different meanings, the word "meaning" itself can in principle have many different meanings. Hence the TalaMind architecture is open to many different ways of representing meanings at the three conceptual levels, for example:

- o Linguistic Level: Linguistic Concept Structures
    - • Concepts represented as sentences in a language of thought
    - • Semantic domains – Collections of sentences about a topic
    - • Mental spaces, conceptual blends
    - • Scenarios for simulation of hypothetical contexts
    - • Grammatical constructions for translation and disambiguation of linguistic meanings
    - • Executable concepts for representing and developing complex behaviors
    - • Finite state automata for representing simple behavioral systems
    - • Formal logic representations, e.g. predicate calculus or conceptual graphs.
- o Archetype Level: Cognitive Concept Structures

---

[19] Vogt (2005) showed that perceptually grounded language games can lead to the emergence of compositional syntax in language evolution. Also see Bachwerk & Vogel (2011) regarding language evolution for coordination of tasks.

- Idealized Cognitive Models (Lakoff, 1987)
- Conceptual Spaces as topological or metric structures (e.g. convex regions) in multiple quality dimensions (Gärdenfors, 2000), with support for prototype effects, similarity detection, etc.
- Radial Categories (Lakoff, 1987)
- Image Schemas (Johnson, 1987; Talmy, 2000)
- Semantic Frames (Fillmore, 1975 *et seq.*) and Conceptual Domains (Lakoff & Johnson, 1980; Langacker, 1987)
- Perceptual Symbols (Barsalou, 1993 *et seq.*)

o   Associative Level: Associations and Data Analysis

- Neural networks (e.g. Hinton, 2006)
- Expressions or data structures induced via machine learning algorithms (e.g. Valiant, 2013)
- Bayesian networks (e.g. Pearl, 1988 *et seq.*)

This is just an illustrative, not exhaustive, list of different ways to represent meanings at different conceptual levels, which will be discussed in greater detail as needed in the following chapters.

So, clearly there is not a consensus view in modern linguistics about how word senses (meanings) exist and should be represented. Indeed, much modern work on computational linguistics is corpus-based and does not directly represent word meanings and definitions. A respected lexicographer wrote a paper (Kilgarriff, 1997) saying he did not believe in word senses. However, Kilgarriff (2007) clarified his position and continued to support research on word sense disambiguation (WSD) (Evans *et al.*, 2016). A sub-community within computational linguistics conducts research on WSD, reported in annual SemEval workshops.

A general view of cognitive semantics[20] is that word senses exist with a radial, prototypical nature; words may develop new meanings over time, and old meanings may be deprecated; words when used often have meanings that are metaphorical or metonymical and may involve mental spaces and conceptual blends[21]; commonsense reasoning and encyclopedic knowledge may be needed for disambiguation

---

[20] See Evans & Green (2006).
[21] See Fauconnier & Turner (2002).

relative to situations in which words are used; the meanings of words and sentences in general depend on the intentions of speakers (viz. Kilgarriff, 2007).

Note that a representation of meaning may span levels and forms of representation, e.g. a linguistic concept structure may reference a cognitive concept structure. Also, some authors may disagree with this placement at different levels. Thus, Fauconnier and Turner might argue mental spaces and conceptual blends should be at the archetype level. While conditional probabilities fit the associative level, Bayesian networks may represent semantics of sentences at the linguistic level in future research. Within the scope of this thesis, precisely how concepts are represented in the archetype and associative levels is not crucial. A Tala agent may not need to include all the different forms of concept representation listed above, particularly at the archetype level, since these overlap in representing concepts. Ways to unify representations within or across the three levels may be a worthwhile topic for future research.

2.2.3   Does Human-Level AI Require Embodiment?

Though the TalaMind approach focuses on the linguistic level of conceptual processing, a Tala agent also includes environment interaction systems with lower levels of conceptual processing, as discussed in §1.5 and shown in Figure 1-1. Consequently a Tala agent can in principle be embodied in a physical environment. So, to the extent that understanding natural language requires embodiment, the TalaMind approach supports this.

However, embodiment does not require that an intelligent system must have physical capabilities exactly matching those of human beings. This would imply that human-level intelligence requires the human physical body and could only be possessed by people. Yet we know people have human-level intelligence even when born without senses like sight or hearing. Also, the unexplained features of human-level intelligence, and in particular the higher-level mentalities, can be described in terms that are essentially independent of the human body (viz. §2.1.2).[22] So, there should be no reason in principle why human-

---

[22] Perhaps the only exception would be the first-person, subjective experience of consciousness. Yet the possibility that other species might possess human-level intelligence suggests that human-level intelligence

level artificial intelligence must require human physical embodiment.

And we should note that embodiment for humans is not what people normally think it to be: We do not have direct knowledge of external reality, or even direct knowledge of our bodies. Instead, we have an internal, *projected reality* (Jackendoff, 1983) constructed from our perceptions of external reality and our perceptions of our bodies. This can be appreciated by considering various illusions, both in our perceptions of external reality and in our perceptions of our bodies (e.g. virtual body illusions). Such illusions show that our perceptions are internal constructs that indirectly represent external reality and our bodies, sometimes incompletely, inaccurately, or paradoxically. It is only because our perceptions generally track reality very closely that we normally think we directly perceive reality.

The TalaMind approach accepts that a language of thought must be embodied by reference to perceptions of an environment, yet that such perceptions are generally incomplete and potentially inaccurate. Understanding of concepts related to the environment, one's body, or physical systems in general can be achieved indirectly by representing knowledge of physical systems and by reasoning within and about such representations. Such reasoning may amount to a mental simulation. A variety of different kinds of representations may be useful, e.g. image schemas, finite state automata for representing behaviors of simple systems, mental spaces, conceptual simulation, etc. These representations may exist within a Tala agent's projected reality or elsewhere in its conceptual framework.

In these pages, this idea is called *virtual embodiment*. It allows an intelligent system to understand and reason about physical reality and to transcend the limitations of its physical body (or lack thereof) in reasoning about the environment – perhaps in the same way a person blind from birth may reason about sight, without direct experience or memory of sight. The projected reality of a TalaMind conceptual framework will be virtual and indirect, though it could in principle be

---

does not require the subjective experience of what it is like to have a human body. Thus it's clear other species (e.g. dolphins, whales, octopi, elephants, ...) have substantial intelligence and yet have very different physical senses and embodiment. And it's at least conceivable that extraterrestrial intelligence may exist comparable or superior to humans, yet with different physical bodies from humans.

interfaced with physical reality (viz. §4.2.2.4).

To appreciate how limited our perceptions are of reality, consider that the frequency of visible light ranges from about 400 to 800 trillion cycles per second, while our nerve cells can only transmit about 1000 pulses per second. So, the reality we see visually is producing waves that oscillate hundreds of billions of times faster than we can perceive. The processing of information by 140 million neurons in each hemisphere's visual cortex, as well as many more neurons elsewhere, enables a 100-billion- neuron human brain to perceive a visual projected reality with great complexity. Yet what we perceive is only a miniscule fraction of the complexity of events around us, happening at different scales of space and time within external reality.

Also, what we perceive of reality is qualitatively different from what actually exists. For example, we now know that what we perceive as solid objects are in reality almost entirely empty space pervaded by force fields and subatomic particles (Close, 2009). So, our human projected reality is inaccurate at the lowest levels of physics, though it is pragmatically very accurate at our day-to-day level of existence.

Our ability to have this knowledge, and to transcend our projected reality, is an example of our own virtual embodiment: It is only by applying human-level intelligence that after generations of thought and experiment we have been able to find ways to virtually perceive aspects of reality that are either too small or too fast for us to perceive in projected reality (such as viruses, microbes, molecules, atoms, subatomic particles, the speed of light, etc.) or too large or too slow for our projected reality (such as Earth's precession about its axis, evolution of species, continental drift, the Sun's lifecycle, the size and age of the universe, etc.)

2.2.4   Natural Language, Metacognition, Inner Speech

Natural language plays an important role in 'broad metacognition' (§2.1.2.6) for human-level intelligence: Mental discourse (inner speech) is perhaps the single best example of broad metacognition involving perception and/or action, as well as reasoning, in human intelligence.

Inner speech is a feature people ascribe to their minds and a psychological phenomenon that has been remarked upon for centuries: We have the ability to mentally hear some of our thoughts expressed internally in natural language. Baars and Gage (2007) write that inner speech is not just for verbal rehearsal but provides an individual's "running commentary" on current issues, and is related to linguistic and semantic long-term memory. Fernyhough (2016 *et seq.*) describes functional MRI studies of inner speech, indicating it can involve parts of the brain that are often used to understand other people's points of view (Theory of Mind or "perspective taking"), and that inner speech may be a conversation between multiple points of view. He suggests inner speech may help our intelligence to be self-directing, flexible, and open-ended.

In terms of broad metacognition, inner speech corresponds to a perception of a silent speech action expressing a thought in natural language. The thought that is expressed may be a phrase, statement, or question about anything, in any domain: The thought may refer to a perception of an external situation or event, or to an (actual or possible) action in the external environment, or it may refer to another thought, or to an emotion, or to oneself, or to a combination. So the thought expressed by an inner speech act may itself indicate further broad metacognition.

These considerations suggest inner speech is not an epiphenomenon, but may play a role in human intelligence, and that natural language may play a role in representing thoughts within the mind, beyond its role for communicating thoughts between people. The fact that we hear inner speech suggests some thoughts are represented internally in a language of thought with the expressiveness of natural language.[23]

So, the fact that we hear inner speech supports the cognitive plausibility of a natural language of thought within a computer model of cognition. Fernyhough's studies also indicate the cognitive plausibility of using a natural language of thought in a 'generalized society of mind' architecture, as described by Doyle (1983) rather than Minsky (1986) – viz. §2.3.3.2.

A computer model could (in effect) emulate perception of internal speech acts by pattern-matching list-structures representing syntax of expressions in a natural language of thought, the approach taken in this thesis (cf. Jackson, 2018d). Internal speech acts and mental percepts of them are represented as conceptual expressions in the TalaMind prototype demonstration system (§6.3.6).

---

[23] It has also been reported that deaf people may experience "inner sign language" (Sacks, 1989).

2.3 Relation of Thesis Approach to Previous Research

2.3.1 Formal, Logical Approaches

As noted in §1.1, one broad stream of research related to understanding intelligence has focused on formal logical approaches to representation and processing. If one accepts the Physical Symbol System Hypothesis (§1.4.4), then one may argue these approaches have, in principle, the ability to support intelligent systems, based on their generality for representing symbolic systems. So this thesis accepts the potential value of formal logic approaches and acknowledges that much has been accomplished with them. Further, the TalaMind architecture is open to use of formal, logical systems within it, including systems based on predicate calculus, conceptual graphs, etc.

Thus, we note in particular the work of Hobbs *et al.* (1993 *et seq.*) regarding *interpretation as abduction* in understanding natural language using first-order predicate calculus; the work of Sowa (1984 *et seq.*) and others on conceptual graph structures; and McCarthy's papers on artificial intelligence cited in the Bibliography, as research directions to consider in future extensions of the TalaMind approach.

Hobbs (2004) advocates abduction (reasoning to determine best explanations) to support commonsense, nonmonotonic reasoning for a language of thought. Hobbs *et al.* (1993) discuss how abduction with first-order logic can be used to solve a variety of problems in natural language understanding, including reference resolution, ambiguity resolution, metonymy resolution, and recognizing discourse structure. Hobbs (2004) discusses how it can be used to recognize a speaker's plan or intentions.

Wilks *et al.* (1996b) note that abduction as a form of logical proof is not sufficient for semantic interpretation: given a false premise, one can prove anything, so abduction needs to filter out false hypotheses. Abduction needs to be guided by meta-knowledge and meta-reasoning to determine which hypotheses are most relevant. Together with Hobbs, their remarks show the importance of viewing abduction as providing explanations, rather than just logical proofs – a perspective consistent with Peirce's view of abduction and with Wittgenstein's view of meaning as involving explanations (viz. §2.2.2).

Hobbs (2004) noted that to support commonsense reasoning a language of thought should be able to represent conjunctions, inference, contradictions, predications, and variable bindings, and these features would give the language of thought the expressive power of first-order logic. He noted higher-order logics can be recast into first-order logic using reification (Hobbs, 2003). In addition, he noted the logic for a language of thought must be nonmonotonic: It must be possible for us to change what we believe to be the truth value of a statement if we gain more information.

However, Hobbs (2004) only claimed a language of thought must have at least these features. While these features are necessary, it does not appear they are sufficient for a language of thought.

Formal, logical approaches do not seem to easily provide the broad range of representations we express with natural language, e.g. features of natural language like self-reference, meta-expressions, metaphor, mental spaces, conceptual blends, idioms, modal verbs, verb aspect and tense, *de dicto* and *de re* expressions, metonymy, anaphora, mutual knowledge, etc. – though in principle each of these features should be possible to represent within formal, logical approaches, and many of them have been investigated. For instance, Vogel (2011) discusses a formal model of first-order belief revision to represent dynamic semantics for metaphors and generic statements. Doyle (1980) described a formal logic approach to reflection and deliberation, discussed further in §2.3.5.

It has been an implicit assumption by AI scientists over the decades that computers must use formal logic languages (or simpler symbolic languages) for internal representation and processing of thoughts in AI systems. It does not appear there is any valid theoretical reason why the syntax and semantics of a natural language like English cannot be used directly by an AI system as its language of thought, without translation into formal languages, to help achieve human-level AI (§3.2.1, §3.3, §4.2.5). There would be theoretical advantages for using a natural language of thought in an AI system: Natural language already has syntax and semantics that can support extensibility, self-reference, meta-reasoning, metaphors, temporal references, analogical reasoning, causal and purposive reasoning, and inference in any domain. Using a natural language of thought would also make an AI system's reasoning more understandable to humans, supporting beneficial AI (§8.2.1).

2.3.2 Cognitive Approaches and Cognitive Linguistics

If formal, logical approaches are one broad stream of research related to understanding intelligence and natural language semantics, then cognitive approaches may be considered as "everything else." This includes a wide variety of approaches by researchers in Artificial Intelligence, Linguistics, Anthropology, Psychology, Neuroscience, Philosophy, and Education.[24]

In AI research, cognitive approaches include Newell's unified theories of cognition, Minsky's society of mind architecture, Wilks' work on preference semantics, Schank's work on narrative case-based dynamic memory structures, Sloman's (1971 *et seq.*) research, Sowa's cognitive architecture, and work by many other researchers on cognitive architectures and neural networks. Some of this research will be discussed in more detail in §2.3.3.

Outside of AI, linguists, psychologists, philosophers, neuroscientists, and researchers in other fields have developed approaches to understanding intelligence and natural language. Many of these researchers would not agree their approaches can be replicated by computers: There is no general agreement among cognitive scientists that human-level AI is possible. Perhaps the only general agreement within cognitive science is that what happens within the human brain cannot be explained simply by observing external behavior, i.e. behaviorist psychology is not sufficient, and one must consider internal information and processes in the brain, to understand the mind.

The TalaMind approach is consistent in many respects with cognitive linguistics research, such as work on Embodied Construction Grammar (ECG) by Feldman (2002 *et seq.*) and Bergen *et al.* (2004), or the research of Steels and de Beule (2006) on Fluid Construction Grammar. ECG provides a computable approach to construction grammar, with embodiment represented via simulation of discrete events. ECG also has grounding in a connectionist, neural theory of language. ECG is relevant to this thesis by providing an existence proof that a computational approach may be considered "embodied." Fluid Construction Grammar research has focused on demonstrating the evolution and emergence of language, using constraint processing for identification and matching in embodied systems, which is an interesting topic for future research in the TalaMind approach, outside the scope of this thesis.

One difference of the TalaMind approach appears to be that previous

---

[24] Fields included within Cognitive Science, listed by the Cognitive Science Society.

approaches do not provide constructions for an internal language of thought. Rather, they provide constructions for external natural language parsing and generation, with internal representations of semantics that in general have been somewhat restricted and apparently not described as languages of thought.

Many researchers in cognitive linguistics have not supported a language of thought hypothesis, but have developed multiple other descriptions for the nature of internal conceptualizations. Some cognitive linguists have expressly rejected a computational language of thought hypothesis.[25] Lakoff (1987, p.343) presented arguments against the viability of an artificial mentalese. However, he left the door open that AI researchers could develop representations that would mesh with his approach to cognitive models.

Evans (2009) presents a cognitive linguistics account of meaning construction in natural language called Lexical Concepts and Cognitive Models (LCCM) theory, which appears to be consistent with the TalaMind approach. He describes lexical concepts as being based on construction grammar, so that by extension it appears his semantic structures can include multi-word expressions, e.g. sentences. He describes LCCM cognitive models as being similar to Barsalou's (1999) description of simulators and perceptual symbols (§4.2.2.4), and as encompassing frames and simulations. Thus, Evans' lexical concepts correspond to the linguistic level of Figure 1-1, and his cognitive models for conceptual structure correspond to elements of the archetype level (although not identical to Lakoff's idealized cognitive models).

LCCM theory is consistent with the TalaMind approach in using conceptual structures based on natural language at the linguistic level, interacting with an archetype level. LCCM theory is different from the TalaMind approach in several respects. For instance, LCCM is not a theory of how to achieve human-level AI; it does not describe a conceptual framework at the linguistic level; it does not include Hypotheses I and III of this thesis; and it does not discuss support of higher-level mentalities.

---

[25] Thus the terms *mentalese* and *language of thought* are not mentioned in either of the comprehensive texts on cognitive linguistics by Evans & Green (2006) or by Croft & Cruse (2004).

2.3.3   Approaches to Human-Level Artificial Intelligence

This section gives some further discussion of research toward human-level AI, augmenting the brief discussion in §§1.1 and 1.2.

2.3.3.1   Sloman

Sloman (1978) published a high-level description of an architecture for an intelligent system that would be able to work flexibly and creatively in multiple domains. He wrote that to achieve artificial intelligence comparable to an adult human, it would be necessary to develop a baby machine that could learn through interaction with others. In general, Sloman's (1978) discussion and some of his subsequent work appear to have been in a similar direction to this thesis, though with different focus. Sloman's (2008) discussion of "generalized languages" for representation is similar though not identical to the TalaMind natural language mentalese hypothesis.

2.3.3.2   Minsky

As noted in §1.5, the TalaMind hypotheses do not require a generalized "society of mind" architecture, but it is consistent with the hypotheses and natural to implement a society of mind at the linguistic level. Since Minsky (1986) described the society of mind as a theory of human-level intelligence, this section provides a brief discussion of his ideas and of similarities and contrasts with the TalaMind approach.

Singh (2003) gave an overview of the history and details of Minsky's theory, noting that Minsky and Papert began work on this idea in the early 1970s. Minsky's description and choice of the term *society of mind* were evocative, inspiring research on cognitive architectures more broadly than he described, to the point that the idea may be considered a paradigm for research. Thus, the term may be used in either of two senses:

1.   The society of mind as proposed by Minsky, including a specific set of methods for organizing mental agents and communicating information, i.e. K-lines, connection lines, nomes, nemes, frames, frame-arrays, transframes, etc.

2.   A society of mind as a multi-agent system, open to methods for organizing agents and communication between agents, other than the methods specified by Minsky, e.g. including languages of thought.

Other sections of this thesis will use the term *society of mind* with the second, *generalized* sense, though not precluding future research on use of Minsky's proposed methods for organization and communication within TalaMind architectures.

To give a few examples of the second perspective, Doyle (1983) described a mathematical framework for specifying the structure of societies of mind having alternative languages of thought. [26] More recently, Wright (2000) discussed the need for an economy of mind in an adaptive, multi-agent society of mind. Bosse and Treur (2006) gave a formal logic discussion of the extent to which collective processes in a multi-agent society can be interpreted as single-agent processes. Shoham and Leyton-Brown (2008) provide an extensive text on multi-agent systems, including a chapter on communication between agents. Sowa (2011) describes communication of conceptual graphs between heterogeneous agents in a framework inspired by Minsky's society of mind.

Minsky described a society of mind as an organization of diverse processes and representations, rejecting the idea that there is a single, uniform process or representation that can achieve human-level intelligence. This thesis is compatible with Minsky's tenet – the TalaMind architecture is envisioned to enable integration of diverse processes and representations.

However, issues related to a language of thought are an area of difference between the TalaMind approach and Minsky's theory. He considered that because agents would be simple and diverse, in general they would not be able to understand a common language. Agents would need different representations and languages, which would tend to be very specialized and limited.

Thus, Minsky did not describe agents in a society of mind sharing an interlingua. He described other, lower-level ways for agents to partially communicate, which he called K-lines and connection lines. To exchange more complex descriptions, Minsky proposed an "inverse-grammar-tactic" mechanism for communication by reconstructing frame representations (viz. Singh, 2003).

In contrast, the TalaMind approach enables agents in a society of

---

[26] As example languages, Doyle discussed logic (FOL – Weyhrauch, 1980), list structures and rational algebraic functions (CONLAN – Sussman & Steele, 1980), and nodes and links (NETL – Fahlman, 1979).

mind to share a language of thought based on the syntax of a natural language.[27] Two agents can communicate to the extent that they can process concepts using common words, and can share pointers to referents and senses of the words. Pattern-matching can be used to enable an agent to recognize concepts it can process, that were created by other agents. This will be discussed and illustrated further in Chapters 3, 5, and 6. An agent in a society of mind may reason directly with concepts expressed in the Tala mentalese, or it may translate to and from other representations and languages, if needed.

Chapter 3 will also discuss how the Tala mentalese can support representing and reasoning with underspecification in natural language. This is compatible with Minsky's (1986, p.207) discussion of ambiguity in thought within a society of mind.

Although Minsky attributed the ambiguity of thought to the act of expression being a process that simplifies descriptions of mental states, the TalaMind approach allows individual thoughts to be ambiguous, just as natural language sentences can be. For instance, in the TalaMind approach the agents in a society of mind could communicate and process the thought *"In most countries most politicians can fool most people on almost every issue most of the time"*[28] (Hobbs, 1983) without needing to consider all the sentence's different logical interpretations, and without needing to consider nonsensical interpretations (viz. §3.6.3.7).

Per §1.6, a society of mind will only be developed in this thesis to a limited extent, as needed to illustrate the thesis approach.

2.3.3.3 McCarthy

Two papers by McCarthy (2007, 2008) considered the general problem of how to achieve human-level artificial intelligence. He said to achieve human-level AI we would need to create systems that can be successful in situations requiring commonsense about information. He said these are situations in which known facts are incomplete; there are

---

[27] This corresponds somewhat to the idea of a "network of question-answerers" described in Jackson (1974, p.328) which suggested a form of emergence for such systems, in the potential for a network of question-answerers to answer a question that could not be answered by a single agent in the system.

[28] Reprinted with permission of Jerry Hobbs and the Association for Computational Linguistics.

no *a priori* limits on what facts are relevant; it cannot be decided in advance what phenomena are to be considered; concepts and theories are approximate and cannot be fully defined; nonmonotonic reasoning is needed to reach conclusions; and introspection may be needed about the system's mental state.

Though McCarthy supported extending mathematical logic formalisms to operate in such commonsense situations, he allowed that some other approach might work. McCarthy (2007) listed several problems that would confront any approach to human-level AI, related to representation of knowledge about the world, nonmonotonic reasoning, reasoning about as well as within contexts, introspection, etc.

Humans have historically used natural language to describe and help solve these problems, and natural language already possesses syntax to represent their semantics. Hence these problems may be plausibly represented and solved within a human-level AI using a mentalese with the expressive scope of natural language, as proposed in this thesis.

McCarthy (2008) discussed the design of a baby machine approach to human-level AI. In general, his discussion is consistent with the approach of this thesis, which would agree the system needs to have an initial set of concepts corresponding to innate knowledge about the world. He lists several kinds of innate conceptual knowledge the system should have, which in general could be supported in the TalaMind architecture. It appears the major difference between McCarthy's perspective and this thesis is regarding the nature of the language of thought that a well-designed baby machine should have. McCarthy wrote that a robot's language of thought should be based on logic, and not on natural language. Responses to his objections are given in §4.2.5.

2.3.3.4   Reverse-Engineering the Brain

Markram (2006) describes the Blue Brain project, for which the long-term goal is to perform detailed, biologically accurate computer simulations of a human brain's neural processing. This approach, reverse-engineering the brain, appears to have the potential to achieve human-level AI. Arguably, the physical processes used by the brain to achieve intelligence could be simulated by computers – especially since, if needed, emerging technologies for computation could be applied, e.g. nanotechnology, quantum computation, etc. However, it is beyond the scope of this thesis to discuss the technical feasibility of this approach.

At minimum the Blue Brain project, and related research, should yield insights into human brain function and could also help support other research toward human-level AI. For example, such research may identify computational neural modules that could be simulated in the associative level of a Tala agent, perhaps supporting Barsalou's perceptual symbols (§4.2.2.4).

2.3.3.5   Cognitive Architectures and AGI

Several authors have conducted research into cognitive architectures and/or *artificial general intelligence* (AGI). This includes Newell and Simon's research (discussed in §2.3.3.6), and research discussed in Albus and Meystel (2001), Anderson and Lebiere (1998), Cassimatis (2002 *et seq.*), Doyle (1980 *et seq.*), Forbus and Hinrichs (2006), Goertzel and Pennachin (2007), Laird, Lebiere and Rosenbloom (2017), Langley, Choi and Rogers (2009), Lenat (1995), Pollock (1990 *et seq.*), Schlenoff *et al.* (2006), Schmidhuber (1987 *et seq.*), Sowa (2011), Swartout *et al.* (2006), and Wang and Goertzel (2012).

Kotseruba and Tsotsos (2018) give an overview of 84 cognitive architectures developed over 40 years of research, of which 49 architectures are presently being actively developed. They report that over 900 practical projects were implemented using these architectures.

In general, these efforts do not discuss research in the same direction as the TalaMind approach, i.e. an intelligence kernel using a language of thought based on natural language syntax and semantics.

Yudkowsky (2007) advocates levels of organization in "deliberative general intelligence" (DGI) as a direction for future research in AGI. The DGI paper proposes a research direction somewhat similar to the TalaMind approach, although the DGI and TalaMind approaches were developed independently. The DGI paper does not present a prototype design or demonstration of its proposed approach. It includes a proposal for "Seed AI" that is similar to the TalaMind intelligence kernel hypothesis (§§1.4.1, 2.3.5). DGI's five levels of organization map into the three levels of conceptual processing discussed in §1.5. In particular, the archetype level corresponds to DGI's layer for concepts, and the linguistic level includes DGI's layers for thoughts and deliberation. Yudkowsky's description of the thoughts layer (2007, p.407) is similar to the TalaMind natural language mentalese hypothesis (§1.4.2) and to Evans' LCCM theory (§2.3.2). However, it appears Yudkowsky (2007, pp.458-461) does not expect that DGI thoughts will (at least initially) be represented as sentences in a natural language mentalese, nor does Yudkowsky propose representing thoughts in structures corresponding to parse-trees of natural language expressions, as this thesis discusses in §§3.3, 3.4, 3.5. Also, DGI focuses on mental images for reasoning. To contrast, this thesis focuses on linguistic reasoning, with spatial reasoning and visualization left as topics for future research.

To the extent that DGI envisions internal use of concept structures different from the Tala natural language mentalese, its proposed research direction appears similar to that investigated by Sloman (§2.3.3.1), or to that implemented independently by Sowa's (2011) VivoMind Cognitive Architecture (VCA). Sowa describes VCA as using conceptual graphs for communication within a society of mind architecture (§2.3.3.2), and as a scalable, efficient system supporting applications that include natural language processing.

2.3.3.6 Newell and Simon's Cognitive Research

In the decades after their groundbreaking research on artificial intelligence in the 1950s, Newell and Simon continued their research and wrote a series of papers about cognitive systems. Simon wrote several books, including one with Newell in 1972 on *Human Problem Solving*. Newell wrote a book in 1990 on *Unified Theories of Cognition*. Their 1976 Turing Lecture proposed the Physical Symbol System Hypothesis (viz. §1.4.4).

2.3.3.6.1 Unified Theories of Cognition

Newell (1990) advocated that scientists develop a series of progressively more complete unified theories of cognition. His initial list of areas to eventually be covered by a unified theory included problem solving, perception, language, emotion, imagination, learning, and self-awareness. He noted the list was incomplete and could be expected to grow. He also made clear that unified theories should be simulated by working computer systems. Thus, the broad scope of a unified theory corresponds to achieving human-level artificial intelligence. His advocacy for unified theories of cognition was in itself an important step toward human-level AI.

Newell (1990) intentionally did not focus on language; he gave reasons for discounting Minsky's (1986) 'society of mind' theory; and he noted that Soar did not address consciousness. In contrast, this thesis focuses on support for a natural language of thought, for the axioms of artificial consciousness proposed by Aleksander and Morton (2007), and for generalized societies of mind (Doyle, 1983). However, there do not appear to be any essential conflicts between Newell's (1990) advocacy of unified theories of cognition and the research direction of this thesis.

2.3.3.6.2 The 'Knowledge Level' and 'Intelligence Level'

Newell (1982) proposed the existence of a "knowledge level" for computer systems, above the level of symbolic processing. Jackson (2018c) discussed theoretical and practical faults of Newell's proposed knowledge level: It is unreal, unchangeable, potentially infinite, and unnecessary.

Newell said an agent at the knowledge level is composed of a set of actions, a set of goals, and a body. Knowledge is a "medium" the agent processes. He said there are no laws of composition for these components, and there is an absence of structure at the knowledge level.

Newell did not define "knowledge" for a system at the knowledge level. He said it is whatever an agent uses to choose actions to achieve goals, according to a "principle of rationality." This principle was given a circular definition, referring to the use of knowledge that an action will achieve a goal.

Newell also said an agent at the knowledge level may have infinite knowledge, because an agent knows all the consequences of everything it knows. He said real systems can only approximate the knowledge level. Given these issues, he said intelligent systems cannot be defined entirely in terms of the knowledge level, and that representations need to exist at the symbolic processing level.

Newell (1990) continued to advocate his 1982 definition of the knowledge level. He defined "perfect" intelligence as an agent using all its knowledge to achieve goals, and said thermostats have perfect intelligence, while humans have imperfect intelligence.[29]

All these problems can be avoided by taking a different theoretical approach. To begin, we observe that Newell (1982) gave an insightful

---

[29] Newell's (1990) discussion of "bands of action" was different from the theoretical idea of a potentially infinite knowledge level: The bands of action are based on real processing in finite human brains. Likewise, his description of unified theories of cognition was different from the unrealistic, potentially infinite knowledge level. Confusingly, he intermixed discussion of the knowledge level with these other ideas.

discussion of *computer system levels*, including the electronic device level, circuit level, logic level, register-transfer level, and symbolic program level. [30] Each computer system level is a functional specialization (subset) of the systems that can be described at the next-lower level. That is, each level provides functionality that not all systems at the next-lower level provide. For example, circuits at the logic level perform logical operations, a functionality that not all systems at the circuit level perform.

According to Newell and Simon's (1976) Physical Symbol System Hypothesis, a subset of the systems at the symbolic program level can achieve human-level artificial intelligence. Jackson (2018c) observed that this subset will be a functional specialization of systems at the symbolic program level, and therefore it will be a computer system level above the symbolic program level. Human-level AI will exist at this level, if and when it is achieved.

It is appropriate to call this future, new computer system level the *intelligence level*, or more fully, the *human intelligence level*. Systems at the intelligence level will be real, finite, changeable, useful systems. [31] They will support Newell's unified theories of cognition, discussed in the previous section.

However, an AI skeptic might say the intelligence level is just as unreal as Newell's knowledge level, arguing that human-level AI is impossible or just an undefinable, fictional idea. To address AI skeptics, this thesis gives a proposal for how to define human-level intelligence and how to design and implement systems having human-level artificial intelligence. In describing the intelligence level and claiming it will exist, Jackson (2018c) also described the TalaMind approach. [32] Arguments of AI skeptics are further discussed in §4.1 and §4.2.4. A significant minority of AI experts may be skeptics about human-level AI

---

[30] Newell remarked that the knowledge level broke many of the rules he identified for these real, physical computer system levels.

[31] Although human-level AIs will be finite systems, they will be able to reason about infinity (using finite concepts), the same way that human mathematicians do (cf. §4.1.2.3).

[32] The human brain is an existence proof that systems can exist at the intelligence level, if the brain can in theory be completely simulated by a large enough digital computer. However, an AI skeptic might argue that the brain cannot be simulated by a computer.

(viz. §8.2.18).

Jackson (2018c) noted that systems at the intelligence level will be able to create, read, and debug computer programs, and to understand how computer programs are supposed to operate for human purposes (in part by reading comments in programs). This could eventually have a real consequence, causing unemployment for human computer programmers. Technological unemployment is discussed in §8.1.

2.3.3.7 Other Influences for Thesis Approach

The approach proposed in this thesis has been influenced by several previous research efforts related to the analysis of human-level intelligence, including Gärdenfors' (2000) discussion of conceptual spaces; Gelernter's (1994) discussion of focus of attention and the "cognitive pulse"; Hofstadter's (1995) discussions of fluid concepts and analogies; and Mandler's (1988 *et seq.*) study of how babies develop an extensible representation system with conceptual primitives.

2.3.4 Approaches to Artificial Consciousness

Blackmore (2011, pp.286-301) gives an overview of research on artificial consciousness. Much of this research has derived from work in robotics and has focused on the associative level of conceptual processing (viz. Figure 1-1).

Aleksander (1996) writes that in 1991 he began investigating artificial consciousness based on neural nets. He and Morton (2007) propose five *"axioms of being conscious,"* using introspective statements:

1. I feel as if I am at the focus of an out there world.
2. I can recall and imagine experiences of feeling in an out there world.
3. My experiences in 2 are dictated by attention, and attention is involved in recall.
4. I can imagine several ways of acting in the future.
5. I can evaluate emotionally ways of acting into the future in order to act in some purposive way.[33]

Aleksander uses first-person statements to address Chalmers' (1995) "Hard Problem" of explaining the subjective experience of

---

[33] Reprinted with permission of Igor Aleksander, Helen Morton, and Imprint Academic. Earlier versions of these axioms were given by Aleksander & Dunmall (2003) and Aleksander (2005).

consciousness (cf. Aleksander, 2005, pp.156-158).

This approach supports at least one answer to the Hard Problem, namely that if the subjective, first-person aspect of consciousness is an illusion, then in principle machines could also have this illusion (viz. Blackmore 2011, p.285). Of course, we are not interested in machines simply giving canned responses saying they believe they are conscious; we want to be able to point to the internal design of the machine and the processing within it that supports a machine's having perceptions of itself, developing beliefs, and acting as if it believes it is conscious (viz. §2.1.2.8). Section 4.2.7 will discuss the relationship of the Hard Problem to the TalaMind approach.

Aleksander and Morton's five axioms may be taken as theoretical requirements for the TalaMind architecture to demonstrate aspects of consciousness, discussed further in §§3.7.6 and 4.2.7, though this thesis intentionally omits discussion of emotion in relation to consciousness and does not focus on attention in recall; these are topics for future research. In addition, reflective observation is included in the list of theoretical requirements for TalaMind to demonstrate consciousness, which seems to be implicit in Aleksander's discussions.

Aleksander and Morton (2007) discuss "depictive architectures" to satisfy these axioms, focusing on the kernel architecture proposed by Aleksander (2005). They define a depiction as "a state in a system S that represents as accurately as required by the purposes of S the world, from a virtual point of view within S" and describe kernel architectures in terms of neural state machines. This is analogous to the TalaMind approach, which §3.7.6 discusses at the linguistic concept level, while depictive architectures are discussed at the associative concept level (viz. Figure 1-1).

Aleksander (1996, 2001) accepts Searle's arguments against symbolic AI, and does not appear to allow his approach to go beyond the associative level of concept processing. This thesis leverages Gärdenfors' (1995) discussion of three levels of inductive inference (§1.5) and does not accept Searle's arguments, in agreement with Chalmers as well as with many AI researchers (viz. §4.2.4).

Sun (1997 *et seq.*) describes research on learning and artificial consciousness, representing explicit knowledge via symbolic rules and implicit knowledge via neural networks. Symbolic rules can be extracted from neural networks and selected via hypothesis testing, to support learning. He gives experimental results on performance of the approach in learning tasks such as the Tower of Hanoi, artificial grammar learning, process control, and minefield navigation.

Chella *et al.* (1997 *et seq.*) discuss the integration of three levels of concept representation to support artificial consciousness, including symbolic concepts expressed as semantic networks and cognitive concepts represented via conceptual spaces (Gärdenfors, 2000), with expectations at the linguistic level helping to guide recognition at lower levels. This is consistent with the TalaMind approach.

Rosenthal's (2005) theory of consciousness in terms of "higher-order thoughts" is synergistic with the TalaMind approach, though he discounts the value of using natural language as a representation for internal thoughts, claiming human thoughts usually do not need to address fine distinctions in meaning that are intrinsic in natural language. The use of natural language syntax in the Tala conceptual language greatly facilitates expression of higher-order thoughts, since it allows Tala conceptual sentences to include other sentences, nested to an arbitrary degree. The use of the reserved variable ?self in TalaMind appears equivalent to Rosenthal's discussion of the need for a first-person indexical in higher-order thoughts. Investigation of Rosenthal's theory within the TalaMind approach would be an interesting topic for future work.

2.3.5   Approaches to Reflection and Self-Programming

Another perspective on artificial intelligence, related to artificial consciousness, is given by research on the topics of reflective and self-programming systems. It is an old, but as yet unrealized and still largely unexplored idea that computer programs should be able to extend and modify themselves, to achieve human-level AI.

In this thesis, self-programming is proposed by the intelligence kernel hypothesis (§1.4.1), which is a variant of Newell and Simon's (1976) Physical Symbol System Hypothesis (§1.4.4). Other authors have proposed similar ideas: Schmidhuber (1987 *et seq.*) investigated self-referential, self-improving systems. Nilsson (2005) [34] proposed that human-level AI may need to be developed as a "core" system able to extend itself when immersed in an appropriate environment, and wrote

---

[34] Nilsson cited a private communication from Ben Wegbreit, ca. 1998, and the 1999 version of McCarthy's *The well-designed child*, cited here as McCarthy (2008).

that similar approaches were suggested by Wegbreit, Brooks (1997), McCarthy, and Hawkins and Blakeslee (2004). Yudkowsky (2007) proposed creating "seed AI" systems that could understand and improve themselves recursively. In 2011, papers by Goertzel, Hall, Leijnen, Pissanetzky, Skaba, and Wang were presented at a workshop on self-programming in AGI systems. Thórisson (2012) discusses a "constructivist AI" approach toward developing self-organizing architectures and self-generated code. Coincidentally, the prototype TalaMind demonstration system illustrates some of the architectural principles Thórisson advocates (e.g. temporal grounding, self-modeling, and pan-architectural pattern-matching), at least to a limited degree (§§5.4.14, 5.4.9, 5.5.3).

Doyle (1980) discussed how a system could defeasibly perform causal and purposive reasoning to reflectively modify its actions and reasoning. He described a conceptual language based on a variant of predicate calculus, in which theories could refer to theories as objects, and in which some concepts could be interpreted as programs. Doyle noted that the use of predicate calculus was not essential, but did not discuss a language of thought based on the syntax of a natural language. His thesis did not include a prototype demonstration, though elements of the approach were partially implemented by himself and others. He expected the approach would require much larger computers than those available in 1980. The TalaMind approach is compatible with Doyle's thesis. The following chapters explore similar ideas to a limited extent, as a subset of the TalaMind architecture.

Smith (1982) studied "how a computational system can be constructed to reason effectively and consequentially about its own inference processes."[35] Though he focused on a limited aspect of this problem (procedural reflection, allowing programs to access and manipulate descriptions of their operations and structures), he gave remarks relevant to human-level AI. He stated the following *"Knowledge Representation Hypothesis"* as descriptive of most AI research at the time:

> "Any mechanically embodied intelligent process will be comprised of structural ingredients that a) we as external observers naturally take to represent a propositional account of

---

[35] Quotations in this section from Smith (1982) are used with permission of Brian C. Smith and MIT Press.

the knowledge that the overall process exhibits, and b) independent of such external semantical attribution, play a formal but causal and essential role in engendering the behavior that manifests that knowledge."

This may be considered as a variant of PSSH (§1.4.4), and describes much AI research up to the present. It is consistent with Hypothesis I of this thesis, to the extent that concepts are considered as symbolic structures (expressions) that represent knowledge. Though in the TalaMind approach, conceptual structures may also represent questions, hypotheses, procedures, etc., each of these may be considered a form of knowledge. Thus, a question may be considered as a statement that something is not known.

Smith provided the following *"Reflection Hypothesis"* as a statement guiding his research into self-reflective systems:

> "In as much as a computational process can be constructed to reason about an external world in virtue of comprising an ingredient process (interpreter) formally manipulating representations of that world, so too a computational process could be made to reason about itself in virtue of comprising an ingredient process (interpreter) manipulating representations of its own operations and structures."

This is also consistent with PSSH, and with Hypothesis I of this thesis. Thus, Hypothesis I may be seen as combining Smith's two hypotheses into a single statement.

Smith gave general remarks about reflection and representation, which are consistent with the TalaMind approach and architecture. More specifically, he wrote:

> "The successful development of a general reflective calculus based on the knowledge representation hypothesis will depend on the prior solution of three problems:
>
> 1. The provision of a computationally tractable and epistemologically adequate descriptive language,
>
> 2. The formulation of a unified theory of computation and representation, and
>
> 3. The demonstration of how a computational system can reason effectively and consequentially about its own

inference processes."

Smith did not pursue the first problem, "in part because it is so ill-constrained." This thesis adopts Hypothesis II, within the TalaMind architecture, to investigate the first problem.

Regarding the second problem, Smith wrote that "every representation system proposed to date exemplifies what we may call a dual-calculus approach: a procedural calculus...is conjoined with a declarative formalism (an encoding of predicate logic, frames, etc.)." He suggested "this dual-calculus style is unnecessary and indicative of serious shortcomings in our conception of the representational endeavor." However, he wrote "this issue too will be largely ignored" in his thesis.

In developing Hypotheses I and II within the TalaMind architecture, this thesis investigates a unified approach to the second problem: The Tala conceptual language provides a representation for both declarative and procedural knowledge, based on the syntax of a natural language.

Smith's thesis focused on the third problem he identified, discussing a limited aspect of this problem. He translated the higher-level problem of how a system could reason about its inference processes into a lower-level problem, i.e. how a programming language could support procedural reflection, allowing programs to access and manipulate descriptions of their operations and control structures, dynamically affecting their interpretation at runtime. This implicitly connects procedural reflection with a form of self-programming. Smith showed how procedural reflection could be incorporated into a variant of Lisp, to support continuations with a variable number of arguments, improve support of macros, etc. Coven (1991) gave further discussion of reflection within functional programming languages, toward support of systems that could in principle reflect on their own reasoning processes and learning algorithms.

Effective reflection and self-programming in human-level AI require computers to have what Smith called "semantic originality" (in other literature called "original intentionality"), i.e. to be able to attribute meaning to symbols and processes independently of human observation. Smith (1982) noted that computers could not yet attribute meaning to what they do, but suggested the possibility they could do so in principle. Haugeland (1985) discussed the topic and its philosophical history at some length, and left open the possibility that computers could in principle attribute meaning. Dretske (1985) discussed requirements for computers to ascribe meaning. Dennett (1987) argued that humans have no greater semantic originality than computers do in principle, because we are biomolecular machines constructed by evolution. Searle (1992) argued that computers cannot in principle attribute semantics – Chalmers (1996) refutes Searle's argument. Section 3.7.2.2 discusses how Tala agents can have semantic originality.

2.3.6   Johnson-Laird's Mental Models

Johnson-Laird (1983 *et seq.*) gives an insightful discussion of many topics relevant to the TalaMind thesis. His 1983 book (here referenced as "MM") discussed three major forms of mental representations, which he called mental models, images, and propositional representations. His theory of mental models continues to be a topic of active research.

Mental models are structural representations of situations, events, and processes in the world.[36] An image is a mental perception of a model from a point of view. Propositional representations are mental representations that correspond most broadly to expressions in natural language (MM, p.165).

MM (pp.410-429) notes that mental models can have different forms and purposes. Broadly, mental models are "iconic" – their structures correspond to structures of situations they represent. Beyond that, mental models may be more or less elaborate, depending on what needs to be represented – a typology includes simple relations, spatial, temporal, kinematic, and dynamic models. Mental models can support spatial-temporal reasoning, which has been previously noted as an important topic for future research to develop human-level AI, outside the scope of this thesis.

The mental models theory stipulates that natural language expressions are represented by propositions in a mental language, which are mapped into mental models (MM, p.165). Johnson-Laird refers to the mental language as a "propositional language," though his discussion throughout MM shows clearly that the language exceeds the semantics of ordinary propositional logic, and even first-order logic. He found that no theory of syllogistic inference satisfies descriptive and

---

[36] Johnson-Laird notes that K. J. W. Craik hypothesized the mind creates such models. Craik (1943, p.83) discussed a "thought-model" that parallels external reality to predict alternative possible events.

explanatory criteria for mental models (MM, p.93).

It seems clear that the TalaMind approach and Tala language (using English as a natural language of thought) are compatible with a mental models approach, which would use iconic mental models of real and imaginary worlds to represent the semantics of Tala expressions (cf. MM, 155-156). Some further references to this are given in §3.6.7.1, §3.6.7.5, and §3.6.7.6.

In discussing how people reason, Johnson-Laird (MM, p.29) found there was no convincing evidence to say that people use any particular logic that corresponds to a formal, mathematical logic. Rather, he showed there are cases where the content of a problem or the way in which it is expressed affects how well people reason about it. In building mental models, people find it easier to represent what is true rather than what is false, which can lead to predictable errors in reasoning. Psychological testing has confirmed such predictions, supporting the theory of mental models (Johnson-Laird, 2010).

Of course, there is a downside to building AI systems that match results of psychological tests by recreating human errors in logic: We don't want to use or rely on systems that can make logical errors. So there are reasons why we should not be solely guided by matching human cognition.

A TalaMind system could have a design for mental models that would not have the potential for logical errors, given consistent premises. It could also be designed to simulate and predict typical errors in human reasoning, using the human-like mental models studied by Johnson-Laird. In principle, this capability could help a TalaMind system identify when problem statements may be confusing to people, and help the system restate problems to avoid confusion.

Johnson-Laird (MM, pp.426-427) notes that mental models can be "meta-linguistic," i.e. contain tokens representing linguistic expressions, and that mental models can be embedded within mental models (MM, pp.430-433). The TalaMind approach also allows inclusion of natural language expressions (represented by Tala structures) within mental models (contexts), to represent what actors within a model may think or say, and nested contexts to represent what an actor may think or perceive other actors think or perceive (i.e. 'theory of mind' capability).

Johnson-Laird (MM, pp.448-477) discussed how a system could have a form of consciousness that (it appears) would support the axioms of artificial consciousness proposed by Aleksander and Morton (§2.3.4,

§3.7.6). Johnson-Laird reasoned that such a system would need to be a parallel processing system, and that a form of self-awareness could result from the system's being able to recursively represent mental models within mental models and to have a higher-level model of its operating system. [37] These features would also be possible in the TalaMind approach.

2.3.7   Research on Natural Logic

Natural logic research[38] has studied how natural language syntax can be analyzed directly to support logically valid reasoning. It is an exception allowed by §1.2's statement that almost all AI research on natural language understanding has attempted to translate natural language into a formal language and perform reasoning with expressions in the formal language. It does not appear that natural logic research has studied use of natural language syntax in a language of thought for an AI system.

2.3.7.1   Natural Logic According to Lakoff

Lakoff (1970) defined a "natural logic" as a logic that would account for inferences made in natural language, and in which non-synonymous sentences would have different logical forms. (This is a summary of five goals he stated for natural logic.) He argued such a logic would need to satisfy a "generative semantics" hypothesis that grammatical rules relate sentence surface forms to logical forms represented using phrase structure trees. Thus, he argued for an approach consistent with Hypothesis II of this thesis, and consistent with the design of the Tala mentalese, which were developed without recalling his paper.

Lakoff wrote that words used in logical forms need additional axioms or "meaning postulates" to characterize their interrelationships and provide models in terms of which logical forms can represent meanings. This corresponds to the ability of words and expressions at the TalaMind linguistic level to refer to concepts and encyclopedic

---

[37] Johnson-Laird (MM, pp.471-477) argued that no Turing machine could be conscious, because consciousness requires a parallel algorithm. Yet he appeared to allow that a sufficiently fast parallel processing system running the right algorithm could be conscious.

[38] Distinct from 'natural deduction', a proof-theory approach to logic (Prawitz, 1965).

knowledge at the archetype level, or to perceptions at the associative level of Figure 1-1.

However, there are several points of difference between Lakoff (1970) and this thesis. For instance, Lakoff's arguments were based on a linguistic analysis of grammaticality and logical sense or nonsense for various English sentences. In contrast, per §3.4.1 the Tala syntax is not limited to expressions people might consider perfectly grammatical English. Per §1.2, this thesis advances Hypothesis II from a strategic perspective, not based on arguments about natural language grammaticality and logical sense or nonsense.

Lakoff (1970) did not discuss many of the topics related to the design of Tala and the TalaMind approach to be discussed in Chapters 3, 5, and 6. His paper was concerned with issues related to human logic and natural language, not with artificial intelligence. Indeed, as §2.3.2 notes, Lakoff has been very skeptical about the prospects for human-level AI. It is not clear he would agree the TalaMind approach can succeed, even though it incorporates ideas from cognitive linguistics, supports embodiment, etc. Lakoff (1970) was also very cautious about prospects for developing a natural logic comprehensive for English grammar, suggesting this would take centuries, if possible at all. However, §5.3 notes that a comprehensive syntax for English is not required for TalaMind's success.

Likewise, this thesis need not fully subscribe to the generative semantics hypothesis, nor to other hypotheses in Lakoff's 1970 paper. Though Tala has a generative grammar, it also supports composable constructions that can transform Tala sentences, effectively extending the grammar. And per §3.2.1 the TalaMind approach is open to use of formal languages such as predicate calculus and conceptual graphs to support understanding natural language and logical reasoning in general. TalaMind does not require that the only logical forms used to represent semantics be phrase structure trees.

2.3.7.2 Monotonicity-Based Natural Logic

Van Benthem (2008) gives an overview of the history of natural logic research. He describes the theoretical background starting with Montague's (1973) analysis of natural language quantifiers. This was followed by analysis of generalized quantifiers (Barwise & Cooper, 1981); analysis of monotonicity for generalized quantifiers (van Benthem, 1986 *et seq.*; Sánchez-Valencia, 1991); and more recently, analysis of semantic relations for containment and exclusion (MacCartney & Manning, 2008 *et seq.*), to give a few highlights of this research. Systems leveraging these theoretical analyses will here be called *monotonicity-based natural logic* systems.

Such systems can compute many entailments of natural language sentences by analyzing parse trees for natural language sentences. Monotonicity-based natural logic has been somewhat successful, in comparison with other approaches to natural language understanding. MacCartney (2009) reports that a natural logic system called NatLog achieved 70% accuracy (and 89% precision) on a subset of the FraCaS test suite[39]containing 183 single-premise natural language entailment problems. On the RTE3 test suite[40] of 800 problems, NatLog achieved 59% accuracy and 70% precision. In comparison, MacCartney notes that a system based on first-order logic (Bos & Markert, 2006) achieved 76% precision on the RTE2 test suite[41] but could only answer about 4% of the problems. NatLog achieved 70% precision answering about 25% of the RTE2 problems.

However, monotonicity-based natural logic has had several limitations. MacCartney (2009) writes that NatLog cannot combine information from multiple premises, and this is a limitation for all other natural logic systems of which he is aware.[42] Because NatLog has a weaker proof theory than first-order logic, it cannot perform some inferences, such as those involving De Morgan's laws for quantifiers, e.g. "Not all x is y ⇔ Some x is not y." MacCartney and Manning (2008) note that many types of inference are not addressed by natural logic, listing examples such as paraphrase, verb alteration, relation extraction, and commonsense reasoning.

In contrast, the TalaMind approach does not have these limitations,

---

[39] Viz. Cooper *et al.* (1996).

[40] Viz. Giampiccolo *et al.* (2007).

[41] Viz. Bar-Haim *et al.* (2006).

[42] It does not appear that supporting multiple premises is impossible in principle for monotonicity-based natural logic. Thus, van Benthem (2008) gives an example involving multiple premises to illustrate how anaphora resolution can be important for monotonicity inferences. MacCartney (2009) notes multiple premises can be supported if combined in a single sentence.

though it could also be considered a kind of natural logic since it involves reasoning directly with natural language syntax. In the TalaMind approach, Tala sentences can use pattern-matching to perform inference with multiple premises; perform translations supporting De Morgan's laws for quantifiers; support paraphrase, verb alteration, and relation extraction; and perform commonsense reasoning, if supported by encyclopedic knowledge (§3.6.7.4) in the TalaMind architecture. (Schubert [2013] discusses how natural logic–like inference can be performed by a formal logic system, EL/EPILOG, also without such limitations.)

The TalaMind approach is open to use of monotonicity-based natural logic, as it is to formal logic methods. The success of monotonicity-based natural logic supports the plausibility of Hypothesis II. TalaMind may provide the "surfacy" natural logic sought by van Benthem (2008).

2.4 Summary

This chapter discussed the relation of the TalaMind hypotheses to previous research, and presented the approach of this thesis to verifying whether a system achieves human-level AI. This approach (design inspection for support of higher-level mentalities) is different from previous research focused on behavioristic comparisons, e.g. via the Turing Test. It is also different from research that seeks to achieve human-level AI through general methods without specifically addressing higher-level mentalities. This chapter's review of previous research has not found an equivalent discussion of the TalaMind hypotheses as a combined approach.

3. Analysis of Thesis Approach to Human-Level AI

A new, a vast, and a powerful language is developed for the future use of analysis, in which to wield its truths so that these may become of more speedy and accurate practical application for the purposes of mankind than the means hitherto in our possession have rendered possible.

~ Ada Lovelace, *Notes*, 1843[43]

∞

3.1 Overview

Chapter 1 presented three hypotheses to address the open question:

How could a system be designed to achieve human-level artificial intelligence?

This chapter will analyze theoretical questions for the hypotheses, and discuss how a system could in principle be designed according to the hypotheses, to achieve the higher-level mentalities of human-level AI. This discussion will use elements of the TalaMind architecture to help answer theoretical questions, and discuss theoretical design issues for elements of the architecture, focusing in particular on the Tala conceptual language. (Appendix A gives a list of theoretical questions considered in this chapter.)

Per §1.6, the analysis presented in this chapter cannot say completely how the proposed architecture should be designed to achieve human-level AI. In general, it can only present theoretical discussions of requirements, design, and feasibility for elements of the architecture. Chapters 5 and 6 discuss what has been done relative to these elements in a prototype demonstration system. Some elements of the design of the prototype system will be used to illustrate the thesis approach, but this chapter is not about the design of the demonstration system, *per se*. It is about more general, theoretical issues, which would apply to any

---

[43] From *Notes upon the Memoir by the Translator*, written by Augusta Ada King, Countess of Lovelace, for her translation of *Sketch of the Analytical Engine Invented by Charles Babbage* written by Luigi Frederico Menabrea of Turin, Officer of the Military Engineers, Bibliothèque Universelle de Genève, October, 1842, No. 82

system built according to the TalaMind hypotheses.

For instance, one of the theoretical issues to be considered is how to represent natural language syntax in a conceptual language based on English. Another set of issues involves how to determine and represent different interpretations and implications of a sentence in the conceptual language. A related set of theoretical issues involves how to represent contexts, and what types of contexts need to be represented in the TalaMind architecture. For each of the higher-level mentalities, we shall need to consider how it can be supported by the proposed architecture, at least in principle theoretically. Such issues will be considered in this chapter as constructive questions, with answers that comprise theoretical arguments in favor of the thesis approach, while Chapter 4 will address theoretical objections to the approach.

3.2 Theoretical Requirements for TalaMind Architecture

This section considers theoretical questions about requirements implied by the hypotheses presented in Chapter 1 for the Tala conceptual language, conceptual framework and processes, to achieve human-level AI.

3.2.1 Conceptual Language

? What is required for a conceptual language to serve as a 'language of thought' for a system with human-level artificial intelligence?

This thesis defines the term *language of thought* for an AI system as a language of symbolic expressions comprising conceptual structures that the system can develop or process. Arguably, a human-level AI system must be able to develop and process conceptual structures that correspond to any linguistically expressible thoughts that a human being can have: If there were some such thoughts that a human being could have, for which an artificial system could not develop and process corresponding conceptual structures, then these thoughts would comprise a realm of thinking beyond the capabilities of the artificial system, and it would not have human-level AI. Therefore it will be taken as a general principle that for a conceptual language to serve as a language of thought for a system with human-level AI, it should include expressions that can represent (correspond to) any human linguistically expressible thought.

Note that we are here making a distinction between thoughts and emotions or sensations. While it is important for a human-level AI system to have some understanding of human emotions or sensations, from the perspective of this thesis it is not required that a human-level AI system be able to experience them (cf. §§2.1.2.9 and 2.2.3). A thought or statement that someone has an emotion is of course different from experiencing the emotion.

? What is the relation of thoughts expressible in natural language to the range of thoughts that need to be expressible in the Tala conceptual language, to achieve human-level AI?

It follows from the preceding answer that the range of thoughts that need to be expressible in the Tala conceptual language includes the thoughts that can be expressed in natural language, since the thoughts a human being can have include those expressible in natural language.

A human-level AI will need to represent other kinds of thoughts, which cannot be easily expressed in natural language. Below the linguistic conceptual level, the TalaMind architecture includes non-linguistic levels of concept representation (Figure 1-1). The topic of whether and how the linguistic level may support concepts not easily expressible in natural language is discussed later. Here it is emphasized that the range of thoughts expressible in natural language is extremely broad, and includes statements about statements, and statements about theories or models, to support meta-reasoning. This implies a language of thought should be a language at a higher level than first-order logic (cf. §2.3.1).

? What properties must the Tala conceptual language have, to represent concepts that can create and modify concepts, to behave intelligently in an environment?

This question is motivated by considering Hypothesis I in conjunction with the definition that Tala as a language of thought provides linguistic representation of concepts that a TalaMind system can develop or process.

There are many concepts that may be said to create and modify concepts. A simple example is a rule of inference, which is a concept that in effect creates a concept, given other concepts as premises. More generally in accordance with Hypothesis I, this thesis will consider concepts that describe processes for creating and modifying concepts.

A conceptual language for a system with human-level AI must be able to represent concepts that describe processes, since these are important concepts that people use natural language to communicate in describing how to perform actions, achieve goals, etc. Intelligent systems need to be able to communicate and follow process descriptions that include sequences of steps to perform, conditional performance of steps, and conditional iteration of steps. Thus, Tala as a conceptual language has an implied requirement to be as expressive in describing processes as a universal programming language,[44] though there is no requirement that it be a conventional programming language. Tala must be able to represent concepts that describe how to perform at least simple processes, with descriptions that are based very largely on the syntax of a natural language. This thesis calls concepts that describe how to perform processes 'executable concepts' or 'xconcepts'.

Of course, people do more than just communicate and follow process descriptions: We think about how to change and improve processes, and communicate about this. Thus a conceptual language for a system with human-level AI must be able to represent concepts that describe how to modify processes, including in principle executable concepts that describe how to modify executable concepts. Again by Hypothesis II, this should be based very largely on the syntax of a natural language. The TalaMind demonstration system will illustrate this ability, in a story simulation where a Tala agent reasons about how to change its process for making bread.

Executing executable concepts requires an interpreter process, which by the definition of conceptual processes in §1.5 is effectively a primitive conceptual process in the TalaMind architecture. The set of conceptual processes may be extended by defining executable concepts, but some conceptual processes may also be defined or emergent at lower levels of processing.

---

[44] To be universal (able to describe any process that could be performed by a Turing machine) a programming language need provide only three basic control structure mechanisms: 1) Sequential execution of one statement followed by another; 2) Conditional execution of one statement or another, based on the value of a Boolean variable; 3) Iterative execution of statements, until a Boolean variable is true (Bohm & Jacopini, 1966). Viz. §5.5.2.

? What other properties must the Tala conceptual language have to support human-level artificial intelligence?

In addressing theoretical questions related to support for higher-level mentalities, the following pages will identify other properties Tala should have. As a starting point, these include the properties proposed by McCarthy in 1955:

- Tala will enable a Tala agent to refer to itself and formulate statements regarding its progress in solving problems.

- The Tala conceptual language will enable expression of rules of conjecture, and the TalaMind architecture will support hypothetical reasoning.

- Tala will be as concise as English, because its sentences will be isomorphic to English sentences.

- The Tala conceptual language will enable expression of concepts involving physical objects, events, etc.

- Tala will have the same extensibility as English, in allowing other languages to be defined within it, and used as appropriate.

- The design of Tala will permit extensions to represent mathematical expressions, and to support mathematical arguments.

McCarthy (1980) proposed circumscription as a rule of conjecture to address the qualification problem in representing commonsense knowledge. This topic will be discussed below in connection with representation of problem contexts (§3.6.7.11).

? To what extent might a conceptual language need to go beyond the syntax of a natural language?

People have invented other languages and notations to represent concepts in some domains more concisely and clearly than is possible in natural language. A simple example is any notation or diagram that depicts a relationship that exists at certain points of an array, and not at others. This may be the best way to concisely and precisely describe a situation summarized by a sentence like "Five Englishmen talked with seven Frenchmen", if not every Englishman talked with every Frenchman. In general, representation of spatial concepts is facilitated by maps, diagrams, and images – "One picture is worth a thousand words". Predicate calculus, conceptual graphs, and mathematical notations are other examples of languages outside the syntax of natural language, which could be worthwhile as alternatives or extensions for the Tala conceptual language (viz. §4.2.2.3).

A conceptual language may need to go beyond the syntax of a natural language by enabling semantic annotation of expressions (Bunt, 2007 *et seq.*) to support conceptual processing. Semantic annotation may be supported within the conceptual language itself, or by integrating other, formal languages for this purpose. This topic will be further discussed below.

Natural language includes the ability to extend itself, since it includes expressions of the form "X means Y", where X and Y may be words or syntactic forms. Thus, per Hypothesis III, Tala includes grammatical constructions. In principle, Tala should have the same extensibility as English, to support definition of new languages.

3.2.2   Conceptual Framework

? What capabilities must the TalaMind conceptual framework have to support achieving human-level AI, according to the TalaMind hypotheses?

Section 1.5 defined a TalaMind conceptual framework as "an information architecture for managing an extensible collection of concepts, expressed in Tala". The term 'information architecture' is used as a general, technology-independent description. The TalaMind approach does not prescribe any particular implementation technologies. The term 'managing' means storing, retrieving, and if necessary deleting concepts.

To support achieving human-level AI according to the TalaMind hypotheses, the following are implied theoretical requirements for capabilities to be provided by a conceptual framework:

- *Manage concepts representing current thoughts*. Since Tala is the language of thought in the TalaMind architecture (viz. §3.2.1), the conceptual framework has an implied requirement to support storing and retrieving thoughts represented as sentences in Tala.

- *Manage concepts representing definitions of words*. Since Tala as a language of thought is based on the syntax of a natural language (English), its expressions use English words. The conceptual framework has an implied requirement to support storing and retrieving definitions of words, represented as sentences in Tala. Section 3.6.1 discusses theoretical requirements for the Tala lexicon. One such requirement is to be integrated with encyclopedic knowledge, discussed below.

- *Manage executable concepts, for conceptual processing.* Theoretical requirements for executable concepts were discussed in §3.2.1. Since the TalaMind architecture must support representing and executing executable concepts, the conceptual framework has an implied requirement to support storing and retrieving them.

- *Manage concepts for a perceived / projected reality.* As discussed in §2.2.3, a Tala agent must have concepts representing its perceptions of the current state of its environment. Following Jackendoff (1983), this set of concepts is called 'projected (or perceived) reality'. The conceptual framework has an implied requirement to support storing and retrieving concepts (percepts) from a projected / perceived reality. As discussed in §1.5, this thesis stipulates that percepts at the linguistic level in the TalaMind conceptual framework are represented as Tala sentences, provided via a conceptual interface by lower levels of conceptual processing that interact with the environment (viz. Figure 1-1).

- *Manage concepts for an 'event-memory' of previous events.* As discussed in §2.3.5, Smith (1982) noted that one of the requirements for reflective learning (a higher-level mentality, per §2.1.2.5) is "the ability to recall memories of a world experienced in the past and of one's participation in that world". Therefore, the conceptual framework has an implied requirement to support storing and retrieving concepts representing such memories. We shall refer to this set of concepts as an 'event memory'. This thesis will stipulate that the event memory in the conceptual framework is a record of previous states of perceived reality, including a Tala agent's percepts and effepts within its environment, expressed as Tala sentences. The event memory will also include a record of a Tala agent's thoughts relative to previous states of reality, so that an agent can create reflective thoughts in forms such as "When X happened, I thought Y and did Z".

- *Manage concepts for encyclopedic and commonsense knowledge.* It is accepted in cognitive semantics that understanding the meanings of words depends on encyclopedic knowledge about how they are used in social interactions and in describing the world we experience. Evans and Green (2006, p.206) call this a central assumption of cognitive semantics, but it is more than an assumption; it is accepted based on arguments and evidence adduced by researchers. Here it will be called *the principle of encyclopedic semantics.* Hence, the conceptual framework has an implied requirement to store and retrieve encyclopedic knowledge, integrated with the Tala lexicon. This thesis will stipulate that encyclopedic knowledge at the linguistic level in the conceptual framework is represented as a collection of Tala sentences, not precluding other languages and notations (per §3.2.1) and not precluding representations at lower levels of Figure 1-1. Encyclopedic knowledge is further discussed in §3.6.7.4.

- *Manage contexts of concepts.* The meanings of natural language sentences depend on the contexts in which they occur. This thesis will stipulate that at the linguistic level of concept representation, contexts can be represented by collections of Tala sentences. The conceptual framework therefore has an implied requirement to manage contexts. Section 3.6.6 discusses the role of contexts in semantic inference. Section 3.6.7 discusses different types of contexts that are needed to support higher-level mentalities, and includes perceived reality, event memory, and encyclopedic and commonsense knowledge as types of contexts.

3.2.3   Conceptual Processes

? What capabilities must the TalaMind conceptual processes have to support achieving human-level AI, according to the TalaMind hypotheses?

Section 1.5 defined TalaMind conceptual processes as "An extensible system of processes that operate on concepts in the conceptual framework, to produce intelligent behaviors and new concepts." A general requirement assumed by the TalaMind hypotheses is that the potential scope of conceptual processes is computationally universal. That is, the scope should be equivalent to any processes that can be performed on symbolic expressions by a universal computer. This follows from §1.4.4's discussion that the TalaMind hypotheses are essentially consistent with the Physical Symbol System Hypothesis, since Newell and Simon (1976) defined physical symbol systems as being realized by universal computers.

In analyzing theoretical questions related to the Tala conceptual language, and to support for higher-level mentalities, the following pages will identify specific capabilities to be provided by conceptual processes. As noted in §§3.2.1 and 1.5, the set of conceptual processes may be extended by defining executable concepts, but some conceptual processes may also be defined or emergent at lower levels of processing.

3.3 Representing Meaning with Natural Language Syntax

? Is it theoretically possible to use the syntax of a natural language to represent meaning in a conceptual language?

Though it might be considered obvious and a truism that syntax can represent semantics, there have been contrary philosophical arguments. Chapter 4 responds to such arguments. This section argues that it is theoretically possible to use the syntax of a natural language to represent meaning as expressed in natural language, at least for the purposes of a conceptual language at the linguistic level of the TalaMind architecture.

When people communicate using natural language, they exchange syntactic information, which they use to help understand intended meanings. People also rely on shared knowledge of word meanings, and shared commonsense and encyclopedic knowledge, much of which is also communicated in natural language syntax. Thus, natural language syntax is used frequently to represent natural language meanings by humans.[45]

The fact that the reader can understand these sentences proves this

---

[45] This may be augmented by information in the form of physical gestures. Indeed, physical gestures may convey all the syntactic information when communicating in sign language for the deaf, another form of natural language.

assertion. The only information available to represent the semantics of these sentences is provided by their syntax, and by shared knowledge of word meanings, shared knowledge of commonsense reasoning, and encyclopedic knowledge. There is no direct physical knowledge available to support understanding these sentences, no use of physical gestures to support them.

This argument does not claim or imply that all knowledge necessary to understand natural language semantics can be represented by natural language syntax, nor that all concepts and semantics can be represented using natural language syntax. Figure 1-1 shows that concepts may be represented at other conceptual levels than the linguistic level. A system that does not have human embodiment may at best have limited, indirect understanding of such concepts through virtual embodiment (§2.2.3).

This argument also does not claim or imply that all concepts at the linguistic level can be represented using natural language syntax. Hypothesis II only posits that the Tala conceptual language may be based "very largely" on natural language syntax, yet allows that other kinds of syntax may be needed for some concepts at the linguistic level. This is further discussed below.

The point remains that natural language syntax is used frequently to represent natural language semantics by humans. It is therefore at least theoretically possible to use the syntax of a natural language to represent natural language semantics in a conceptual language at the linguistic level of the TalaMind architecture.

> ? Is it theoretically possible to reason directly with natural language syntax?

There is no reason in principle why inference cannot be performed with a conceptual language based on natural language syntax. Chapters 5 and 6 will present a demonstration system to support this claim. Following is an argument that this is theoretically possible:

When inference is performed with a formal language not based on natural language syntax, such as predicate calculus, the syntax of the formal language is needed to support the operations of inference. It is syntax that enables taking what might otherwise be considered a random string of symbols, and recognizing clauses, variables, logical operators, etc. It is syntax that enables matching these elements in sentences in predicate calculus, to perform an inference and construct an implied sentence in predicate calculus.

Prior to the invention of formal languages, people for millennia used natural language to support inference. Just as they use natural language syntax to represent the semantics of natural language in communication, people also use the syntax of natural language to support inference during communication (apart from whatever mentalese syntax may be used internally for inference). Formal languages such as predicate calculus and existential graphs were originally developed to support inferences that had previously been expressed in natural language (Sowa, 2007a).

A major strength of natural language is its provision of syntax that can support multi-level reasoning in human-level AI, i.e. meta-reasoning, analogical reasoning, and causal and purposive reasoning, as well as abduction, induction, and deduction, about any domain or across domains.

? Is it theoretically valid to choose English as a basis for the Tala conceptual language, rather than other natural languages?

As discussed above, the range of thoughts that need to be expressible in the Tala conceptual language includes the thoughts that can be expressed in natural language. Almost all natural languages have equivalent scope in being able to express human thoughts, though some concepts may be easier to express in some languages than in others (Pinker, 1994). Therefore it is theoretically valid to choose English as a basis for the Tala conceptual language, and it would be valid to choose a different natural language.

There are some reasons for choosing English. In particular, its syntax has been thoroughly studied and documented, perhaps more than any other natural language (Hudson, 2010).

? Which elements of English syntax are important to Tala? What about morphology and phonology?

From the perspective of Hypothesis II, the elements of natural language syntax that are important are those that help to represent semantics. In English, virtually every element of sentence and lexeme syntax supports representing some aspect of semantics.

Also, from a practical standpoint, all elements of natural language syntax may need to be represented in a Tala conceptual language, for any natural language that is used as its basis, to support generating output that appears fluent to humans.

Per §1.4.2, morphology and phonology are topics intended for future research, outside the scope of this thesis. It appears they should be possible to include within the TalaMind approach, using representation techniques similar to those described for sentential syntax in this thesis.

3.4   Representing English Syntax in Tala

This section will present arguments for certain theoretical principles to follow in developing the design of Tala. Then a design for Tala will be described according to these principles, not precluding other designs. Chapter 5 will present details of the design developed for the prototype demonstration system, as a work in progress.

3.4.1   Non-Prescriptive, Open, Flexible

A Tala agent should be able to understand and communicate with natural language as it is actually used by people. The Tala language should therefore be able to represent how sentences are actually expressed, so that meta-reasoning can occur about the way they are expressed.

For example, Sag *et al.* (2003, p.37) list the following as grammatical and non-grammatical sentences[46]:

> The defendant denied the accusation.
> The problem disappeared.
> *The defendant denied.
> *The teacher disappeared the problem.

A traditional prescriptive grammar might not permit the two starred expressions, by specifying that *deny* requires an object while *disappear* does not permit one. Yet a Tala agent could encounter both in the real world, and should be able to process them in contexts such as:

> Question: Did the defendant admit or deny the accusation?
> Answer: The defendant denied.

> A child speaking: The magician disappeared the rabbit!

It sometimes seems that almost every example that may be cited as ungrammatical could happen in real-world usage, whether expressed by children or adults learning a language, by people speaking

---

[46] Reprinted with permission of CSLI Publications.

colloquially or poetically, by people writing concise notes, etc. The Tala syntax should therefore be non-prescriptive, open, and flexible. The syntax presented in Chapter 5 will use a small set of production rules, which describe the structure of Tala sentences, not limited to expressions people might consider perfectly grammatical English. For instance, the syntax will allow that a verb may or may not have a subject and/or object. Conceptual processing relative to the context for use of a verb will be responsible for understanding an expected though missing, or unexpected yet present, subject or object. (A subject for a verb is normally missing in an imperative sentence, and could be missing in a concise expression such as "Looks good.") So, a Tala sentence can express an utterance that may not be a grammatical English sentence.

3.4.2 Semantic and Ontological Neutrality and Generality

To support representing the broad range of human thoughts, the syntax for Tala should, as much as possible, be semantically and ontologically neutral and general: The syntax should minimize assumptions and implications about the concepts that may be expressed in the Tala language. There should be minimal built-in dependencies or restrictions to any predefined ontology, or to any predefined encyclopedic knowledge. Instead, the Tala language should be flexible and powerful enough that sentences in Tala can be used to describe any particular ontology, or set of semantic domains for any particular concepts.

To a large extent this design principle is a given, and something to be preserved in basing Tala on natural language syntax, which does not imply which words should be used to describe reality, in general. The only exceptions are those built into a natural language and its syntax: For instance, English has some built-in ontological preconceptions about time for expression of verb tense and aspect, about point of view reflected in first-, second-, and third-person pronouns, etc.

3.5 Choices and Methods for Representing English Syntax

Having identified certain theoretical principles, we next discuss specific choices and methods for representing English syntax in Tala according to these principles.

3.5.1 Theoretical Approach to Represent English Syntax

? Which theoretical option is chosen for representing English syntax in Tala?

This thesis follows a dependency grammar approach to English syntax: In representing the syntax of an English sentence, the corresponding Tala sentence will show syntactic relations between words, and will not distinguish words from phrase structures. The English syntax for Tala presented in Chapter 5 is similar in spirit to Hudson's 'Word Grammar' for English. Some differences in detail will be noted, as well as opportunities for future development of Tala.

It should be noted that the use of dependency grammar is not required by the TalaMind approach. Other options for representing English syntax are consistent with Hypothesis II and could be explored in future research.

3.5.2   Representing Syntactic Structure of NL Sentences

Another principle in designing the syntax of Tala is that sentences in Tala should describe the syntactic structure of corresponding natural language sentences. The reason for this is that the TalaMind approach seeks to emulate the ability of human-level intelligence to use the syntax of individual natural language sentences for representing and reasoning about their meaning. To support this, information about the syntax of individual natural language sentences needs to be represented in Tala conceptual structures. We shall refer to this as the *structurality principle* for the Tala language.

?   How can the syntactic structure of individual natural language sentences be represented in Tala sentences, to support reasoning with syntactic structures?

This is a method-oriented question, though here it is analyzed from a theoretical perspective to identify a viable approach. We first note that whatever method is selected, it will involve symbolic expressions representing the syntactic structure of English sentences, given that we are concerned with physical symbol systems, i.e. the TalaMind approach is a variant of the Physical Symbol System Hypothesis, as discussed in §1.4.4. By PSSH, the essence of symbolic processing is to manipulate symbolic expressions, to modify, delete, or create new expressions.

A dependency diagram of the syntax of an English sentence is essentially a tree structure, though re-entrant links or lattice structure may be needed (cf. triangular dependencies in Hudson, 2010). Therefore, to represent a dependency diagram of the syntax of an

English sentence as a symbolic expression, one needs a symbolic representation of tree structures that can also support re-entrant links and lattice structures. This symbolic representation needs to be flexible in allowing syntactic information to be at each node. And this representation needs to be accessible in data structures for symbolic processing within some programming environment, to enable construction of a TalaMind architecture that performs conceptual processing of Tala sentences.

Hierarchically nested list structures are a natural choice for all these purposes, supported by symbolic processing environments in Lisp dialects (McCarthy, 1960). Pointers can be used to represent re-entrant links and lattices in list structures. Therefore, Tala sentences are expressed as hierarchically nested list structures, representing the syntactic structure of natural language sentences according to a dependency grammar of English. This format supports reasoning, and conceptual processing in general, in Lisp-dialect symbolic programming environments. Following is an example of a Tala expression for the natural language sentence *"Leo asks Ben can you turn grain into fare for people?"*:

```
(ask
    (wusage verb)
    (subj Leo)
    (indirect-obj Ben)
    (obj
        (turn
            (wusage verb)
            (modal can)
            (sentence-class question)
            (subj you)
            (obj
                (grain
                    (wusage noun)
                    ))
            (into
                (fare (wusage noun)
                    (for
                        (people (wusage noun))
                        )))))
    (tense present)
    (subj-person third-singular)]
```

For sake of uniformity, each level of the list structure starts with a symbol, followed by an association list. The symbol that starts each level alternates between an English word and a syntactic keyword (`obj`, `subj`, …), with prepositions treated as keywords.

It may be noted that the choice of list structures and Lisp is not required by the TalaMind hypotheses. Other representations for dependency tree structures could be appropriate in different programming environments, e.g. some might prefer XML. A specific design choice for Tala is needed within this thesis to present examples for theoretical discussions, and to support the demonstration system. There are strong technical arguments in favor of list structures and Lisp. And since Tala responds to McCarthy's 1955 proposal for an artificial language corresponding to English, it is altogether fitting to choose Lisp for Tala.

In addition to structural relationships within sentences, the syntax of English includes several words that in conceptual processing effectively function as logical operators, variables, or constants, within or between sentences. These amount to primitive words in the syntax of English, and include prepositions, conjunctions, pronouns, determiners, and certain nouns, verbs, adjectives, and adverbs. They will be discussed further in §§3.6.8 and 5.3.

3.6 Semantic Representation and Processing

3.6.1 Lexemes, Senses, Referents, and Variables

> ? How can a Tala agent represent the different possible meanings of an English word?

According to cognitive lexical semantics, the possible meanings of a commonly used English word are not bounded by a finite list of precise definitions (cf. §2.2.2). Rather, most commonly used words have many different meanings that may at best be approximately described by a list of definitions, and more adequately represented by cognitive constructs such as radial categories (viz. Evans & Green, 2006, p.328).

The architecture shown in Figure 1-1 includes two non-linguistic levels to represent concepts structured according to typicality effects, relative to prototypes. This thesis recognizes their importance, but the specifics of such representations are not central to this thesis. Rather, the TalaMind approach is open to alternatives and future research regarding them.

Although the possible meanings of common words cannot be completely enumerated, definitions are nonetheless useful in constructing meaning. And if a human-level AI is to have semantic originality, then it must be able to express definitions, to declaratively represent that "X means Y" (viz. §§2.3.5, 3.7.2.2). Hence, one of the roles of the TalaMind conceptual framework is to include concepts representing definitions of words. Per Hypothesis II, these concepts should be expressed as sentences in the Tala conceptual language, based on the syntax of natural language, though it should be possible to have pointers from Tala sentences to concepts represented in the non-linguistic levels.

Although this thesis refers to the Tala lexicon as if it were a distinct part of the conceptual framework, by the principle of encyclopedic semantics (§3.2.2) the lexicon should be integrated with encyclopedic knowledge, also to be stored in the conceptual framework. Encyclopedic knowledge is further discussed in §§3.6.7.4 and 3.6.7.7.

The TalaMind hypotheses do not prescribe any particular approach to structuring the Tala lexicon within the conceptual framework. A natural approach would be to structure the Tala lexicon as a network or lattice, with each node representing a possible meaning of a word and containing information expressed in Tala sentences. This could correspond to network and inheritance approaches described by Daelemans *et al.* (1992 *et seq.*), Tyler and Evans (2003), Hudson (2007 *et seq.*), and other researchers.[47]

> ? How can a Tala agent represent the specific senses of the words in a given sentence?

A Tala sentence uses words as both lexemes and word senses, distinguishing lexemes with the `wusage` syntax field. To represent the sense(s) in which a word is used, a `wsense` syntax field can point to one or more meanings in the Tala lexicon, represented as natural language sentences or phrases, or to concepts represented non-linguistically. If the lexicon is structured as a network of concepts, then the `wsense` field could contain pointers to nodes in the network. Such notations are not problematic from a theoretical perspective, so this thesis will not define a notation for pointers into the lexicon, nor for that matter pointers to concepts in the non-linguistic levels of Figure 1-1. However, a pointer notation will be introduced later for use within Tala sentences, to support representing various kinds of interpretations.

---

[47] WordNet (Miller, 1991) would be a natural resource to leverage in building a Tala lexicon. Automatic parsing of WordNet definitions could be used to generate definitions expressed in Tala.

As will be discussed in §3.6.3, the `wsense` field can also be used for semantic annotation within Tala sentences, e.g. to indicate that "an" is used in a sentence with a specific or non-specific sense. The senses defined in the lexicon can help construct intended meanings, but do not prohibit other meanings, since in general word senses cannot be finitely enumerated, and can be underspecified. And of course, people often convey meaning without adhering to definitions.

? How can a Tala agent represent the specific referents of words in a given sentence?

Though the `wusage` and `wsense` fields allow representing that the lexeme 'bat' is used as a noun with the sense "nocturnal flying mouselike mammal", the question remains how to represent that it refers to a specific bat, e.g. one observed flying in the evening.

To represent the referent(s) of a word in a sentence, a `wreferent` syntax field can point to particular concepts in the Tala conceptual framework, or to concepts represented non-linguistically. Within the linguistic level of Figure 1-1, one of the roles of the conceptual framework is to include concepts representing percepts. Such concepts may also be represented at the non-linguistic levels. The `wreferent` syntax field enables words in Tala sentences to point to percepts, with the pointer notation syntax being implementation-dependent.

The `wreferent` field can contain a pointer to another Tala expression to support coreference within Tala sentences, e.g. to specify referents of anaphora (§§3.6.3.12, 3.6.8), "shell nouns" (Schmid, 1997 *et seq.*), etc. Section 3.6.6.2 gives an example of a `wreferent` pointer for the shell noun "statement" in representing the Liar's Paradox.

? How can a Tala agent determine the specific senses and referents of the words in a given sentence?

Per §1.5's description of the TalaMind architecture, there are logically three ways that a Tala sentence can originate within a conceptual framework. First, it can be an innate sentence, created when the architecture is constructed. Second, it can be created as a result of conceptual processing of other Tala sentences that exist in a conceptual framework. Third, it may be received via the conceptual interface from environment interaction systems. In general, all these methods may be expected to provide Tala sentences that already specify word usages, senses, and referents, with one important exception: Environment interaction systems may provide Tala words or sentences originating in English spoken or written by people, which may be incomplete in specifying usages, senses, and referents.

Natural language parsing to determine word usages may occur in the environment interaction layers of Figure 1-1, or as conceptual processing within the TalaMind architecture, or both. The syntax of Tala is a reasonable target for natural language parsing, but developing parsing approaches is a topic for future research, per §1.6.

Per §1.5, it is the role of conceptual processing to determine how to elaborate Tala words or sentences to specify senses and referents of words, and even word usages if not determined by parsing. If a Tala agent receives the sentence "There is a bat" in its conceptual framework as a spoken utterance via the conceptual interface, it is the role of conceptual processing to determine the word sense and referent for "bat". Determining that it refers to a specific flying mammal requires abductive processing of information available in context, within the conceptual framework. This will involve use of percepts, interactive context, and encyclopedic and commonsense knowledge in the conceptual framework (viz. §3.6.7.4). To support abductive conceptual processing, the TalaMind approach and architecture are open to use of computational linguistics methods for word sense disambiguation (viz. Navigli, 2009), though this is a topic for future research.

Prior to determining senses and referents, an English word is essentially a variable in Tala. As these are determined within a context, the variable becomes bound. To the extent its senses and referents are not bound, an English word's presence in a Tala concept is a form of underspecification, akin to a metaconstant or metavariable in a formal language for underspecification (cf. Bunt, 2008). In addition to English words as variables, Tala also provides a separate syntax for variables that can be bound to words and Tala expressions, to support pattern-matching and pointer references within the conceptual framework. There are some additional aspects of this topic related to meanings of primitive words in Tala and English, which will be discussed in §3.6.8, after more groundwork has been established.

3.6.2 Multiple Representations for the Same Concept

? Can there be different Tala sentences that express the same concept?

Yes. Answering this question "No" appears to have motivated much work on previous concept representation approaches. For example, Schank (1973, p.191) stated a precept for his work on conceptual dependency structures, that if two natural language expressions mean the same thing, even if they are in different languages, there should be only one conceptual structure that represents their meaning.

Relative to this question, the only implication of Hypothesis II in this thesis is that every sentence expressible in natural language has a corresponding Tala sentence. Since a concept may be expressed by many natural language sentences, it may also have many Tala sentences that express it. This may be considered a theoretical advantage of the TalaMind approach, because different ways of expressing a concept may express important information, such as focus of attention, or pragmatic information about how the concept was expressed. Representing the precise way a concept was expressed supports meta-reasoning about how it was expressed, in addition to reasoning with the concept itself. It is also a practical advantage, since it avoids significant extra work from researchers striving to agree on a single way to represent each concept expressible in natural language.

3.6.3 Representing Interpretations

? How can a Tala agent represent different possible interpretations of an English sentence?

If a Tala representation of an English sentence has multiple possible interpretations then it follows from Hypothesis II that in general more specific interpretations may be represented by other Tala sentences. Exceptions are allowed per §3.2.1, but this rule handles several problems discussed below.

As noted in §3.2.1, a conceptual language may need to go beyond the syntax of a natural language by enabling semantic annotation of expressions (Bunt, 2007 *et seq.*). Slots such as (wsense), (wreferent), and the (<- ?p) syntax for pointers (which provides a form of coreference representation; viz. §5.3.11) are forms of semantic annotation within Tala. Similarly, quantifiers like 'all' or 'an' can be annotated in Tala by adding (wsense collective) or (wsense specific) slots. Developing a complete syntax for semantic annotations in Tala is a topic for future research, as is potential integration of Tala with other formal languages for semantic annotation.

3.6.3.1 Underspecification

In human conversation it is often not necessary to express a sentence that has only one possible interpretation, and in fact it may be difficult to do so, since most common words have many possible meanings, and most sequences of words can be interpreted in multiple ways. Natural language expressions are often used with ambiguities because the speaker doesn't think at the level of precision corresponding to the disambiguated readings, because the speaker thinks it is immaterial which interpretation is chosen, and/or because the context makes one specific interpretation much more plausible than others.

In the TalaMind architecture, conceptual processing of a Tala sentence can occur at the level of precision in which a sentence is expressed, without using a more precise interpretation. The Tala conceptual language is effectively a language for underspecified semantic representations (cf. Bunt & Black, 2000, p.24) using the syntax of natural language, rather than creating a separate formal language for underspecification. In the TalaMind approach for achieving human-level artificial intelligence, a Tala agent has the same ability and burden for its internal conceptual processing that human intelligence has in external communication, to use ambiguous sentences or to create more specific interpretations when appropriate.

3.6.3.2 Syntactic Elimination of Interpretations

Since a Tala sentence represents a dependency parse-tree of an English sentence, its structure can eliminate some possible interpretations of a linear text sentence. Consider the classic example, "Time flies like an arrow". A Tala conceptual structure for the normal interpretation is:

```
(fly
     (wusage verb)
     (subj
          (time
               (wusage noun)
               ))
     (like
          (wusage prep)
          (arrow
               (wusage noun)
               (det a)
               ))
     (tense present)
     (subj-person third-singular)]
```

This conceptual structure eliminates the interpretation that the sentence is an instruction to measure (time) the speed of flies the same way that one would measure the speed of an arrow, since in that interpretation "time" would be a verb, and "fly" would be a noun.

The conceptual structure still has other ambiguities. For example, the meaning of the adverbial preposition "like" and the verb "fly" may be ambiguous, relative to the noun "time". With further conceptual processing, another conceptual structure could be created to represent a more specific interpretation, such as:

```
"time passing" resembles "an arrow passing".
(resemble
        (wusage verb)
        (subj
                (pass
                        (wusage verb)
                        (aspect continuous)
                        (subj
                                (time
                                        (wusage noun)
                                        ))))
        (obj
                (pass
                        (wusage verb)
                        (aspect continuous)
                        (subj
                                (arrow
                                        (wusage noun)
                                        (det an)
                                        )]
```

At this point the conceptual structure is a more reasonable expression of a normal interpretation of the original sentence. It could be made more specific by identifying senses of its nouns and verbs within the conceptual structure. Further conceptual processing could develop other, related concepts as interpretations or inferences from the original concept.

This approach does not require that a natural language sentence have only one 'deep structure' representation – rather it allows multiple, different structural representations to be developed and elaborated to represent different interpretations, if needed, with each interpretation represented in the Tala language. Variables can be used in a Tala structure to represent unknowns, if needed.[48]

---

[48] Cf. Pinker's (2014) brief, introductory discussion of Chomskyan deep structures.

3.6.3.3   Generic and Non-Generic Interpretations

An example of a sentence that has such interpretations is "The lion is a dangerous animal". Depending on context, the interpretations can be represented by different Tala sentences, e.g. corresponding to "In general, each lion is a dangerous animal" and "The lion in that cage is a dangerous animal".

3.6.3.4   Specific and Non-Specific Interpretations

An example of a sentence that has such interpretations is "Mary wants to marry an Irishman." Depending on context, the interpretations can be represented by different Tala sentences, e.g. corresponding to "Mary wants to marry a specific Irishman" and "Mary wants whomever she marries to be an Irishman, but does not want to marry a specific man." These interpretations could also be represented by semantically annotating "an" in the original sentence with either a (wsense specific) or (wsense non-specific) marker.

It may be noted that "Mary wants to marry a specific Irishman" still has multiple interpretations, according to how the Irishman may be specified. The normal interpretation is that Mary can identify one man whom she wants to marry, who happens to be Irish. Yet a technically valid interpretation would be "Mary wants to marry the specific Irishman who happens to win the next National Lottery of Ireland." In that case she cannot yet identify the particular individual, but it could be argued there is a specific Irishman she wants to marry. The point remains that either interpretation of the original sentence can be expressed in a Tala sentence, if necessary.

3.6.3.5   Individual and Collective Interpretations

An example of a sentence that has such interpretations is "Two men lifted the piano". Depending on context, the interpretations can be represented by different Tala sentences, e.g. corresponding to "Two men working together at the same time lifted the piano" and "Two men lifted the piano, at different times, each acting individually".

3.6.3.6   Count and Mass Interpretations

An example of a sentence that has such interpretations is "There's no chicken in this chicken soup". Depending on context, the interpretations can be represented by different Tala sentences, e.g. corresponding to "There is not an entire chicken bird in this soup made of chicken meat" and "There is not any chicken meat in this soup claimed to be made with chicken meat."

3.6.3.7 Quantificational Interpretations

If a natural language sentence has multiple quantifiers to be interpreted for scope and distributivity, there may be multiple ways they can be interpreted. Some simple examples of such sentences are:

Five Englishmen talked with seven Frenchmen.

Every male student dates an undergrad.

Every representative of a company saw most samples.

Each of the above sentences has only a single, obvious dependency parsing, corresponding to a Tala sentence, but has at least two interpretations:

Five Englishmen talked with seven Frenchmen.

A group of five Englishmen talked with a group of seven Frenchmen.

A total of five Englishmen talked with a total of seven Frenchmen, in unspecified groups.

Every male student dates an undergrad.

Each male student dates an undergrad, who is typically a different person for each male student.

Each male student dates an undergrad, who is the same person for all male students.

Every representative of a company saw most samples.

Every representative of each company saw most samples.

Every representative of one company saw most samples.

In the last example, a normal interpretation of "most samples" is that it varies for different representatives and companies, but technically it could also mean the same majority set of samples was seen by each representative and company. 'Most' can also be interpreted as an aggregate across other variables, e.g. "A minority of taxpayers pay most of the taxes" does not mean that each taxpayer in the minority pays most of the taxes, but that together they do. 'Few', 'some', and 'all' have similar variations in interpretation. A more complex quantified sentence may have many more possible interpretations, and a corresponding predicate calculus expression may become implausibly long and complicated as a meaning representation. For example:

> In most democratic countries most politicians can fool most of the people on almost every issue most of the time.[49] (Hobbs, 1983)

A normal interpretation is that the quantifier 'most' varies across countries, politicians, people, and issues in an unspecified manner that may not be important.[50] Yet technically the sentence has interpretations where the quantifier does not vary. For example, one precise interpretation is that people between the ages of 10 and 80 can be fooled about the same majority set of issues across a certain majority of democratic countries, but only between 2 a.m. and 5 p.m. Such an interpretation is not supported by commonsense, but is logically permitted. A human-level AI should not need to consider it, since it does not affect how humans might discuss Hobbs' example sentence.

Quantified sentences are often cases where natural language enables people to summarize situations imprecisely, leveraging underspecification to support reasoning without considering more complex precise interpretations. Typically what matters when such sentences are used is only a general result or implication, not a specific, precise interpretation. Human intelligence can reason with imprecise, underspecified sentences as if they are true, even if some interpretations are false or nonsensical, or contradictory to other interpretations. Human-level AI must be able to do the same, for people to judge that a system possesses human-level intelligence.

To illustrate, a salesman may want to know that many of his sample products were seen by representatives of many companies at a trade show. He might ask an associate "Were our products seen by many companies at the show?" and receive the answer "Every representative of a company saw most samples."

---

[49] Reprinted with permission of Jerry Hobbs and the Association for Computational Linguistics.

[50] If it is important, then depending on context in discourse, a speaker may wish to claim that particular issues, countries, etc. are exceptions.

```
(see
        (wusage verb)
        (tense past)
        (subj
                (representative
                        (wusage noun)
                        (det every)
                        (of
                                (company
                                        (wusage noun)
                                        (det a)
                                ))))
        (obj
                (sample
                        (wusage noun)
                        (number plural)
                        (det most]
```

He interprets this as an affirmative answer to his question, because in conceptual processing of the Tala or English sentence, it is simple to interpret the determiner 'a' as 'each', yielding "Every representative of each company saw most samples." (This is similar to interpreting "The lion is a dangerous animal" as "In general, each lion is a dangerous animal.") The salesman may not even consider the interpretation "Every representative of one company saw most samples" since he asked a question about many companies, and the reply is easily translated as about each company, which refers to every company individually. The salesman also avoids any need for a precise interpretation of how the quantifier 'most' varies within the sentence, specifying which representatives of which companies saw which samples. If he cares about that relationship, he may ask for it specifically, using this natural language which-expression.

To summarize, Tala enables representation and facilitates conceptual processing of both precise and imprecise quantifications. This is a theoretical advantage of adopting Hypothesis II for a conceptual language, in support of human-level AI, compared with attempting to reason in formal languages such as predicate calculus. It enables Tala to be used as a language for underspecified semantic representations rather than creating a separate, conventional formal language for underspecification.

3.6.3.8 *De Dicto* and *De Re* Interpretations

McCarthy (2008) noted that a language of thought would need to include pointers to support certain kinds of internal mental references. The Tala (<- ?p) pointer-binding syntax for coreference (§5.3.11) can support representing *de re* interpretations. Consider the example:

Columbus believed that Cuba was India.

The *de re* interpretation is that Columbus believed the location now called Cuba was the same as the location called India. This interpretation is essentially correct (Columbus did initially believe he had reached Asia), though Columbus was in error and they are not the same location. It can be represented in Tala as follows:

```
(and
     (denote (wusage verb)
          (subj India)
          (obj (location (wusage noun) (<- ?i)))
          )
     (denote (wusage verb)
          (subj Cuba)
          (obj (location (wusage noun) (<- ?c)))
          )
     (be (wusage verb)
          (adv not)
          (subj ?i)
          (obj ?c)
          )
     (believe (wusage verb)
          (subj Columbus)
          (tense past)
          (aspect perfect)
          (obj
               (be (wusage verb)
                    (tense past)
                    (obj ?i)
                    (subj ?c)
                    ]
```

This uses the syntax (<- ?i) to bind the variable ?i to a pointer referring to the Tala expression in which (<- ?i) occurs. So the Tala sentence says that "India" and "Cuba" denote locations that are not the same, which Columbus believed were the same.

The *de dicto*, literal interpretation, that Columbus thought a location he knew was named Cuba was the same as a location named India, can be represented as follows:

```
(believe (wusage verb)
     (subj Columbus)
     (tense past)
     (aspect perfect)
     (obj
          (be (wusage verb)(tense past)
               (subj
                    (location (wusage noun)
                         (obj-of
                              (denote (wusage verb)
```

```
                    (tense past)
                    (aspect perfect)
                    (subj Cuba)))))
    (obj
      (location (wusage noun)
        (obj-of
          (denote (wusage verb)
          (tense past)
          (aspect perfect)
          (subj India]
```

This interpretation is false, because Columbus did not know that any location was named "Cuba", even though he thought he had reached India. This example just shows that each interpretation can be represented differently in Tala (viz. Rapaport *et al.*, 1997).

3.6.3.9   Interpretations of Compound Noun Structures

An interpretation of a compound noun structure can be represented as a Tala concept structure:

```
"steel ship engine"
(engine
    (wusage noun)
    (obj-of
        (use
            (subj
                (ship
                    (wusage noun)
                    (subj-of
                        (make
                            (wusage verb)
                            (passive)
                            (aspect perfect)
                            (of
                                (steel
                                    (wusage noun)
                                    ]
```

"Steel" may refer to the engine rather than the ship, or to what is transported by the ship rather than its construction. Each of these interpretations can be represented as a Tala concept structure. Some other examples are provided by Bunt and Black (2000, p.8).

This approach extends to interpretations for adjectives and adverbs in compound phrases, allowing more specific representation of semantics than supposing they refer to simple attributes.

"green reporter" – reporter who lacks experience.

"green energy" – energy produced by an ecological process, or energy that when consumed does not harm the ecology.

Similar remarks apply to representing interpretations for verb structures.

3.6.3.10 Interpretations of Metaphors

The sentence "That surgeon is a butcher" has both literal and metaphorical interpretations. These can be represented with other Tala sentences, corresponding to:

> That person is a surgeon and is also a butcher, and has appropriate skills in each profession.

> That person is an incompetent surgeon, whose skills as a surgeon would be more appropriate for a butcher.

While there is no difficulty in representing either interpretation as a Tala sentence, there is a theoretical question regarding how an intelligent system could develop the second, metaphorical interpretation. It is an implication, since inference and encyclopedic knowledge are needed to recognize that describing a surgeon as a butcher is likely pejorative, even though both are skilled professions when considered separately (Grady *et al.*, 1999). This topic is discussed in the subsections on semantic inference (§3.6.6.5) and conceptual blends (§3.6.7.9).

3.6.3.11 Interpretations of Metonyms

A sentence like "The Waldorf salad is waiting for her bill" could have a metonymical interpretation, if spoken in a restaurant:

> The customer who ordered the Waldorf salad is waiting for her bill.

There is no difficulty in representing this interpretation as a Tala sentence, but developing it requires additional knowledge related to the context, so it is an implication. (Cf. Lakoff & Johnson, 1980.) This topic is discussed in the subsections on semantic inference (§3.6.6.5) and conceptual blends (§3.6.7.9).

3.6.3.12 Interpretations of Anaphora

Anaphorical interpretations of pronoun references (e.g. in sentences like "The customer is waiting for her bill") can be represented by pointer values in (wreference) slots on pronouns. These pointer values can be bound using the (<- ?p) notation, either in the same sentence or in others within a context (viz. §5.3.11). In general, semantic inference

(§3.6.6) is required to disambiguate these references.

3.6.3.13 Interpretation of Idioms

Fillmore, Kay, and O'Connor (1988) found that idioms are problematic to interpret for the following reasons: non-conformance to ordinary English grammar (*all of a sudden*); inference does not predict conventional meaning from normal word senses (*kick the bucket, spill the beans*); inference does not predict conventionality (*wide awake* is conventional, but not *wide alert*); inference does not predict pragmatic usage from normal word senses (*How do you do?*); the idiom has an open form, for which inference does not readily predict usage with other words (...*let alone*...).

This thesis adopts the approach Fillmore *et al.* advocated, which is to support interpretation of idioms via grammatical constructions. In a TalaMind architecture, grammatical constructions can be represented by executable concepts written as Tala sentences, called *Tala constructions*; a Tala construction can pattern-match all or part of another Tala sentence X and assert a new Tala sentence Y that is effectively an interpretation of X; the new sentence Y can contain content unrelated to X, and (depending on the idiom) Y can also include information obtained from X; a Tala construction can match a sentence or phrase that is ungrammatical; and Tala constructions are composable, i.e. multiple constructions can perform translations in sequence. These features enable Tala constructions to construct meanings not implied by the expressions they match, a central ability of constructions discussed by Goldberg (1995 *et seq.*).

The TalaMind architecture presented in this thesis does not go so far as to represent all the grammar of Tala via constructions, a direction I expect would be advocated by proponents of construction grammars (cf. Kay & Fillmore, 1999; Goldberg, 1995 *et seq.*; Croft, 2002; Evans, 2009). Per §3.5.1, this is a possible direction for future research. On the other hand, Hudson (2007, pp.153-157) discusses how constructions can be used naturally in combination with a dependency grammar. Section 5.5.4 describes an initial implementation of Tala constructions for the TalaMind prototype demonstration system. Section 3.6.7.9 notes that constructions can be used to automatically translate conventionalized metaphorical expressions, as well as idioms.

3.6.4   Semantic Disambiguation

? How can a Tala agent determine which interpretations of an English sentence are appropriate?

In the TalaMind architecture, conceptual processing is responsible for creating different interpretations of a Tala sentence, and for determining which interpretations are appropriate. It is also responsible for using a Tala sentence without creating other interpretations, if appropriate. Some conceptual processing may be relatively automatic. As discussed above, constructions may be applied to create new interpretations for a Tala sentence.

The most appropriate interpretation of a Tala sentence may be an "indirect meaning" determined by pragmatic reasoning, not equivalent to a precise logical interpretation. An illustration of this was given in §3.6.3.7 for quantifications, where an interpretation involved changing the determiner 'a' to 'each'. As another example, if someone says "Are you listening to this music?", the indirect meaning may be "Can we listen to something else, or turn the music off?" Discussions of pragmatic implicature were given by Grice (1989) and by Sperber and Wilson (1986). Bunt and Black (2000) discuss other examples of pragmatic, indirect interpretations.

In general, determining which interpretations are appropriate involves abductive processing and pragmatic reasoning in contexts, relative to commonsense and encyclopedic knowledge, percepts and memories of events, etc. Thus it involves information in the conceptual framework, and information available from non-linguistic levels in Figure 1-1. The processing involved for disambiguation is in essence the same as determining which implications of a sentence are appropriate. So, method-oriented questions related to disambiguation will be discussed theoretically in §3.6.6 on semantic inference and §3.6.7 on representation of contexts. Chapters 5 and 6 provide examples in the prototype demonstration illustrating how the TalaMind approach can support disambiguation.

3.6.5   Representing Implications

? How can a Tala agent represent different possible implications of an English sentence?

Hypothesis II implies that in general each implication of a Tala sentence may be represented by another Tala sentence. The term 'implication' may be used broadly to refer to any concept that might be created by conceptual processing of one or more Tala sentences. So, in addition to logical consequences this usage includes interpretations of a Tala sentence, indirect meanings, and also questions, goals, [51] new (perhaps domain-specific) rules of inference, meta-statements, and so on, *ad infinitum*. It is straightforward to represent these in natural language, and so to express them as Tala sentences. Some natural language sentences prompt for the creation of contexts, e.g. through expressions like "Once upon a time..." (viz. §3.6.7.8). These may be considered a special case of this rule, since contexts will be represented by collections of Tala sentences, to be discussed in §3.6.7.

3.6.6   Semantic Inference

? How can a Tala agent determine which implications of an English sentence are appropriate?

Considering the broad idea of implication mentioned above, which includes questions, goals, meta-statements, etc., this amounts to a discussion of how higher-level mentalities can be supported   in the TalaMind architecture, which will be taken up in §3.7. There are several natural, lower-level questions about inference in the TalaMind approach, to be considered here.

? How can logical inference be performed using the Tala conceptual language, working directly with natural language syntax?

Tala's use of nested list structures to represent syntactic structures facilitates pattern-matching of Tala sentences, which, combined with a syntax for variables in Tala and representation of inference rules as if-then sentences, enables the mechanics of logical inference. Examples of this will be given in Chapter 6.

---

[51] A goal may be represented by a Tala sentence of the form "I want X", with the reserved variable `?self` used in place of the first-person singular pronoun. Evans (2009) notes that the meaning of *want* depends on its object. In principle a Tala agent can specify X internally well enough to represent the meaning of its goals for its own purposes.

3.6.6.1 Representation of Truth

? How is truth represented in the TalaMind architecture?

This thesis stipulates that Tala sentences will exist within contexts in the TalaMind architecture, and that if a Tala sentence exists within a context, this represents the sentence may be processed as if it is true in that context. For example, it may be used as the premise of an inference rule to create another Tala sentence in the same context.

This thesis stipulates that a context is a collection of Tala sentences. There are a variety of different kinds of contexts to be discussed in more detail in §3.6.7. Here we focus just on the general, theoretical nature of inference within a context. Note that these stipulations are not required by the TalaMind hypotheses, but are given to describe a viable approach for implementing the hypotheses.

3.6.6.2 Negation and Contradictions

? How are negation and falsity represented in the TalaMind architecture?

This thesis stipulates that if a negation of a Tala sentence A exists in a context, this represents that the negation is processed as if it is true in the context, and that A may be considered false in the context, to the extent it is negated (as explained below). If neither A nor a negation of A exists in a context, then neither A nor its negation is explicitly represented as true or false in the context. Of course, there may be other sentences in the context that imply A or a negation of A.

Per §3.4.1, Tala is designed to represent how English sentences are actually expressed by people. Negation in Tala is therefore expressed in the same ways it is expressed in English. Negation occurs in English by various forms of grammatical marking, in which 'not', 'never', 'no', and other negative words are used within phrases, and in which negative morphemes (e.g. 'un', 'dis', 'anti') are used within words.

The extent to which these markings express negation in English is often not as straightforward a matter as negation is in symbolic logic. Their meanings depend on where they occur in larger expressions and contexts, and also depend on social and idiomatic conventions. Conceptual processing is responsible for determining such meanings, using encyclopedic and commonsense knowledge. Thus while 'not' is a primitive word in English, it does not have a fixed, constant interpretation. In this respect it is similar to other primitive words in English (viz. §3.6.8).

Horn (2001, p.xx) writes:

> "Marked negation is not reducible to a truth-functional one-place connective with the familiar truth table associated with logical negation, nor is it definable as a distinct logical operator; it represents, rather, a metalinguistic device for registering an objection to the content or form of a previous utterance (not to a proposition) on any grounds whatever, including the implicatures (conventional and conversational), the grammatical and phonetic form, or the register associated with that utterance."[52]

The complexity of negation is indicated by the fact that Horn's study of it spans 500 pages.

For example, scalar negations are often not treated as they would be in symbolic logic. Thus the sentence "He was standing not six feet from me" has the conventional interpretation "He was standing less than six from me" rather than permitting the subject to be seven feet or further away. Similarly "He is not happy" is conventionally interpreted as meaning the subject is sad, distressed, or perhaps angry, while interpretations such as "He is apathetic / puzzled / philosophical..." are often not considered.

Double negatives are another, classic example of natural language expressions often not interpreted according to symbolic logic, to intensify a single negation rather than reverse it. Famous examples include the catchphrase of the late, much-respected comedian Rodney Dangerfield, and the lyrics of the Rolling Stones' song *"Satisfaction"*.[53] Virtually any number of negatives may be used to intensify negation, and must be understood by a system with human-level AI. Double negatives can also involve problematic scalar negations, e.g. "He is not unhappy" does not necessarily mean "He is happy".

Negation can also be implied by sentences worded without negations, by conventions that do not follow symbolic logic. For example, "If I had won the lottery, I would own a Rolls Royce"

---

[52] Reprinted with permission of Laurence Horn & CSLI Publications.

[53] I don't have no permission to reprint quotations of their double-negative sentences in a commercial book.

conventionally implies one did not win the lottery and does not own a Rolls, though neither of these follow according to symbolic logic. Likewise "Some of my funds are in cash" conventionally implies not all of one's funds are in cash, though this does not follow logically. "That surgeon is a butcher" metaphorically implies someone is not a competent surgeon, even though surgeons and butchers are usually competent when considered separately.

Human-level AI will have the same challenges as human intelligence, in understanding negations expressed in natural language. It is an advantage of the TalaMind approach that negations are expressed within a conceptual language based on the syntax of natural language, in a structure that makes them accessible for conceptual processing based on encyclopedic knowledge, including knowledge of idioms, social conventions, etc. This gives reason to think the challenges of understanding negations are surmountable via the TalaMind approach: In principle, conceptual processing can translate Dangerfield's catchphrase into "I don't receive any respect" or translate "not six feet away" into "probably less than six feet away". Semantic annotation can be used to record decisions to treat negations in a Tala sentence as literal or intensified, etc. Counterfactual implications can be interpreted using mental spaces (Cutrer, 1994; viz. §3.6.7.8). Metaphorical implications can be developed using conceptual blends (§3.6.7.9). Per §3.2.1, symbolic logic can still be used when appropriate. However, per §1.6 this thesis will not attempt to fully design how negations in natural language should be interpreted by a TalaMind architecture to achieve human-level AI – this remains a topic for future research.

?   How does the TalaMind architecture guarantee that if a Tala sentence is true, its negation is false?

In general, there is no such guarantee. Conceptual processing is responsible for detecting that contradictory sentences exist in a context, and resolving inconsistencies, if possible and appropriate. Meta-reasoning may occur about the sentences in a context, and about the consistency, usefulness, etc. of a context.

In some contexts, such as mathematical and logical theories, it is crucial that no contradictions exist, and if a contradiction is deduced, the context is invalidated. In other real-world contexts it is likely contradictions will exist, and they may not be possible to resolve, yet the context remains valid for conceptual processing. Human-level AI has the same problem as human intelligence, to manage contradictions in such contexts.

Suppose a Tala agent encounters Captain Kirk, who says "This statement is false." While this sentence could be interpreted as referring to some other sentence depending on the context, it could also be interpreted as referring to itself. That interpretation could be represented in Tala as:

```
(be  (<- ?x)(wusage verb)
     (adv not)
     (subj
          (statement
               (wusage noun)
               (det this)
               (wreferent ?x)))
     (obj (true (wusage adj)]
```

The Tala agent should not go into an infinite loop processing the Liar's Paradox, emitting sparks and smoke.[54] Its conceptual processing should detect that the interpretation is self-contradictory, and stop processing it. There is no reason in principle why computers cannot use meta-reasoning to recognize and transcend logical paradoxes, in the same way people do.

Individuals may hold conflicting views and say self-contradictory things. Thus, Othello may say "I love Desdemona" and "I do not love Desdemona". If a Tala agent were to encounter Othello making both statements, it should not invalidate its context for reality, nor for Shakespeare's play.[55] Nor should it conclude Othello is completely inconsistent in all his thoughts and will believe anything. Rather, a Tala agent should become circumspect regarding implications it might derive from either statement Othello makes about Desdemona. It could withdraw all such implications from its context for Shakespeare's play, and just wait and see what happens. It could make very limited, tentative predictions about what might happen. It could reason about why Othello has made conflicting statements, and whether both

---

[54] *Star Trek: The Original Series*, "*I, Mudd*", episode 37, 1967, based on a story by Gene Roddenberry. Kirk used different wording.

[55] "Excellent wretch! Perdition catch my soul, but I do love thee! and when I love thee not, Chaos is come again." *Othello*, Act 3, Scene 3, ca. 1603 by William Shakespeare.

statements are true but in different ways or at different times. If it has cultural knowledge about watching plays, then it knows it cannot talk to Othello or Desdemona, nor interrupt the play and annoy the audience.

Per §3.6.3.7, real-world contexts are often different from mathematical logic. Logicians are concerned with determining if a sentence is true for all possible interpretations. Human-level intelligence, on the other hand, is able to use natural language sentences that have many possible interpretations, without considering or evaluating all of them. People frequently use sentences as if they are true, even though there are some interpretations that are false or nonsensical. When people do interpret a sentence, they create a context-dependent interpretation, and only become concerned about other possible interpretations if their initial interpretation appears to be inconsistent within the context. The processing to detect an inconsistency may be general and domain independent, or it may be domain specific. To achieve human-level AI, a Tala agent will need to emulate these abilities, to understand intended interpretations of natural language sentences, without insisting on sentences that do not have possibly false interpretations.

? How can it be determined that two concepts are contradictory?

Disagreements between sentences occur within contexts. It can be relatively straightforward for conceptual processing to recognize contradictions and disagreements between sentences, for relatively clear, simple statements of facts. Yet in general, per the discussion of negation above, it requires commonsense, encyclopedic knowledge, and inference, to determine that two sentences imply a contradiction, and it may not always be possible to recognize one. Human-level AI will have to perform much the same kind of conceptual processing that human intelligence does in relation to negation and contradictions, with similar limitations.

3.6.6.3 Inference with Commonsense

? How does the TalaMind architecture guarantee that, if "John and Mary live in Chicago" is true, the sentence "Mary and John live in Chicago" is also true (both on the individual and on the collective readings)?

Via the representation of commonsense and encyclopedic knowledge, integrated with the Tala lexicon in the conceptual framework, for the words in a particular sentence. For the verb "live", the default commonsense knowledge is that if "A and B live", then they do so both collectively and individually, and that the order of specification is irrelevant. Conceptual processes are responsible for generating entailments of Tala sentences, using the appropriate knowledge in context. For a different verb and object, like "have a baby", the commonsense knowledge in the conceptual framework should be different, specifying that if "A and B have a baby" then they do so collectively, order of specification irrelevant, though only the female does so individually.

3.6.6.4 Paraphrase and Inference

? How can conceptual processing deal with the fact that the same concept may be expressed in different ways?

Leveraging commonsense and encyclopedic knowledge, the TalaMind architecture permits use of constructions and other executable concepts to translate Tala conceptual structures from one form to another, so that different conceptual structures with the same meanings can be processed if Tala conceptual processes need them to be expressed in different ways.

3.6.6.5 Inference for Metaphors and Metonyms

? How can conceptual processing determine the implications of a metaphorical or metonymical expression?

As noted in §§3.6.3.10 and 3.6.3.11, determining the implications of metaphors and metonyms requires inference and encyclopedic or context-specific knowledge. For example, to interpret the metaphor "That surgeon is a butcher" as a pejorative statement, encyclopedic knowledge must be used in the following inferences (Grady *et al.*, 1999):

> Since most people have only one skilled profession, the literal interpretation that the sentence describes someone who is both a skilled surgeon and a skilled butcher is unlikely.

> Therefore it is more likely the sentence uses 'is' metaphorically, saying the surgeon is like a butcher.

The goals and methods of a surgeon are different from those of a butcher. A surgeon having the goals and methods of a butcher would not be a competent surgeon.

Therefore a metaphorical interpretation is "That surgeon is incompetent."

In accordance with Hypothesis III, and without precluding other approaches, this thesis will discuss the use of conceptual blends (Fauconnier & Turner, 2002) for conceptual processing of metaphors. This is discussed in §3.6.7.9, along with discussion of metonyms.

Interpreting metaphors is an extremely important problem for natural language understanding and human-level AI. Metaphoric expressions are pervasive in natural language. Hence a system that is not adept at understanding metaphors will not be considered to have human-level intelligence. The importance of metaphors in reasoning by analogy, and imagination, has been discussed by several cognitive scientists, including Lakoff, Johnson, Turner, and Fauconnier. Per Hypothesis III, mental spaces and conceptual blends support the TalaMind approach for higher-level mentalities, as discussed in §3.7.

3.6.7   Representation of Contexts

?   What is context?

From the perspective of computational pragmatics, context may be construed as all the conditions that affect how language is generated and understood (cf. Bunt, 2000). Due to the centrality of linguistic processing in the TalaMind approach, from its perspective context may be construed as all the conditions that influence human-level artificial intelligence. At the linguistic level of the TalaMind architecture, this amounts to saying context is the totality of Tala sentences that influence the conceptual processing of a Tala agent, per the next section.[56]

3.6.7.1   Dimensions of Context

To be more specific, in discussing computational pragmatics it is helpful to distinguish 'dimensions of context' (Bunt, 2000). These may

---

[56] It is envisioned that contexts in the conceptual framework at the linguistic level will be integrated with iconic mental models for spatial-temporal reasoning (§2.3.6), e.g. for reasoning with perceptions of the external environment provided via the associative level.

be adapted for the TalaMind architecture, as follows:

- **Linguistic context**: Information ranging from definitions in the Tala lexicon to both raw and analyzed linguistic material in current media being processed by a Tala agent, current dialog history, etc.

- **Semantic context**: Information ranging from encyclopedic and commonsense concepts to concepts about topics of current linguistic media being processed by a Tala agent, e.g. current tasks or situations being discussed or considered by the agent.

- **Cognitive context**: Concepts describing a Tala agent's conceptual processing, and describing the conceptual processing of others, e.g. beliefs about beliefs and intentions of others. Also, hypothetical contexts, contexts representing problems and theories, and contexts representing ongoing communications are in this category.

- **Physical and perceptual context**: Concepts ranging from a Tala agent's perceived reality and memory of events, to concepts about the availability of communicative and perceptual channels, the presence and attention of partners in communication, etc.

- **Social context**: Concepts about the social aspects of situations, e.g. communicative rights, obligations, and constraints of participants in dialogs.

For the TalaMind architecture there is some overlap in these dimensions, but it is not necessary that they be completely separate in describing aspects of context. For example, knowledge of semantic domains overlaps linguistic, semantic, and cognitive contexts.

In studies of dialog pragmatics, it is helpful to distinguish global vs. local aspects for each dimension, where global refers to information that tends to remain constant throughout a dialog and affects general characteristics of the dialog, and local refers to information that can be changed through communication in a dialog (Bunt, 1994 *et seq.*). The global vs. local distinction is also very relevant in the TalaMind architecture and will be used in the following pages to characterize contexts, though virtually any kind of information can change for conceptual processing in general: New definitions and new encyclopedic knowledge can be developed, new perceptions of reality, etc.

? What types of contexts should be represented and processed in the TalaMind approach?

Following Hypothesis III this thesis will discuss how mental spaces and conceptual blends may be used as contexts supporting higher-level mentalities. Also, the discussion of theoretical questions in previous pages indicates additional context types should be included in the TalaMind architecture, to represent commonsense and encyclopedic knowledge, theories, perceived reality, event memory, hypothetical contexts, interactive contexts, and problem contexts. Representation of each of these context types will be discussed from a theoretical perspective in the following subsections. Conceptual processing of them will be discussed in §3.7 on theoretical support of higher-level mentalities.

These context types are reasonably well-defined and accepted ideas in cognitive linguistics, cognitive science, and AI, but they do not have definitions with the level of precision found in discussions of formal theories. Hence, this chapter can only present general descriptions of these context types, viewed as collections of Tala sentences, and general discussions of their requirements and feasibility, from a theoretical perspective. What has been implemented relative to these elements in the prototype demonstration system is discussed in Chapters 5 and 6.

Further, if context is construed as all the conditions that influence human-level artificial intelligence, then per §1.6 this thesis cannot claim to identify all the types of contexts that need to be represented and processed to achieve human-level artificial intelligence. The TalaMind approach and architecture are not limited to representing only certain types of contexts, *a priori*.

? How can contexts be represented in the TalaMind approach?

Sowa (2001) discusses several approaches to the formal representation of contexts, which relate mostly to the cognitive dimension but have overlap with other dimensions. These include the possible worlds of Leibniz, Kripke, and Montague; model sets of Hintikka; situation theory of Barwise and Perry; Peircean contexts; McCarthy's formalism for contexts; and Sowa's representation of contexts via conceptual graphs. Sowa notes distinctions between actual and possible propositions and contexts, and that Peirce distinguished between logical possibility (a proposition p is consistent with the facts and not proven false or true), subjective possibility (p is believable and not known to be false), and objective possibility (p is physically possible). Additionally a proposition may be questioned, or permissible. For each of these distinctions there is also a corresponding negation. All of these modalities may be mixed within a context, and in addition agents may have propositional attitudes (intentions) toward contexts and propositions. For example, agents may want, fear, or hope a proposition to be true in a context.

This thesis stipulates that a context is a collection of Tala sentences. Given the universality of natural language, Tala has in principle the ability to express all the distinctions needed to represent actual, possible, and intentional propositions.

?   Do contexts have attributes or features in the TalaMind approach?

Though a context is a collection of sentences, different context types will need to have other features or attributes represented. For instance, per §3.6.6.2 one attribute is whether a context tolerates or does not tolerate contradictory sentences within it, and how conceptual processing should behave if a contradiction is detected. As another example, a context may have a default temporal or spatial location relative to the Tala agent, e.g. a hypothetical context may have a relative temporal location in the future, present, or past. A third example is the *epistemic mode* of a context, an attribute characterizing whether the sentences in a context represent factual knowledge, an imaginary or fictional situation, perceptual concepts, etc. (viz. §§3.6.7.7, 3.6.7.8, 4.2.8). The following pages will give initial discussions of attributes and features as needed within this thesis, for different context types.

?   Do contexts have internal structures in the TalaMind approach?

Though a context is a collection of sentences, it will typically need to have additional structure, depending on the type of context. For instance, an interactive context will have a temporal structure, identifying a sequence of interactions. The following pages will give initial discussions of internal structures as needed within this thesis, for different context types. In future work, contexts may be iconic mental models, and support spatial-temporal reasoning (§2.1.2.10, §2.3.6).

3.6.7.2 Perceived Reality

As discussed in §3.2.2, a Tala agent's conceptual framework has a requirement to include a set of concepts representing its projected / perceived reality, i.e. its perceptions of the current state of its environment expressed in Tala sentences, not precluding representations at lower levels of Figure 1-1. Perceived reality can be referenced in ongoing local, interactive contexts (§3.6.7.5) or in hypothetical contexts (§3.6.7.6) considering possible future scenarios. Since perceived reality is a collection of Tala sentences, it is a context type. As noted previously, in future work the Tala linguistic perceived reality context may be integrated with iconic mental models for spatial-temporal reasoning (§2.3.6) with perceptions of an external environment provided via the associative level.

3.6.7.3 Event Memory

As discussed in §3.2.2, a Tala agent's conceptual framework has a requirement to include a set of concepts representing the agent's memories of past events expressed in Tala sentences, not precluding representations at lower levels of Figure 1-1. Event memory can be referenced in local, interactive contexts (§3.6.7.5) ongoing within perceived reality, or in hypothetical contexts (§3.6.7.6) to consider counterfactual scenarios. Since event memory is a collection of Tala sentences, it is a context type.

3.6.7.4 Encyclopedic and Commonsense Knowledge

As discussed in §3.2.2, a Tala agent's conceptual framework has a requirement to include concepts comprising encyclopedic and commonsense knowledge, expressed as a collection of Tala sentences, not precluding representations at lower levels of Figure 1-1. Encyclopedic and commonsense knowledge is effectively a global context, which may be referenced and applicable in other, local contexts. Section 3.6.7.7 discusses semantic domains for access and organization of encyclopedic and commonsense knowledge. For concision, in general this thesis will use 'encyclopedic knowledge' to include commonsense knowledge.

More specifically, the scope of this knowledge includes:

- Propositional knowledge typically found in human encyclopedias and dictionaries, about any topic.

- Sociocultural knowledge about interactional norms and goals in different kinds of situations.

- Commonsense knowledge, considered so widely known it is often not written down.

- Perceptual and behavioral knowledge, needed to recognize or perform what words describe.

As we proceed down this list, the extent to which knowledge can be expressed using natural language diminishes, and these forms of knowledge are increasingly represented at lower levels of Figure 1-1. Thus, encyclopedic knowledge spans all three conceptual levels in Figure 1-1. This list agrees with a description of encyclopedic knowledge given by Evans (2009).

Per §1.6, the topic of encyclopedic knowledge is outside the scope of this thesis, and intended for future research. It is claimed, by virtue of the ontological universality of natural language as a basis for Tala and the stipulated computational universality of TalaMind conceptual processes, that encyclopedic knowledge can be represented and processed in the TalaMind architecture – particularly since the lower levels of Figure 1-1 offer generality in representing concepts not easily expressed in natural language. This is a strong claim in the sense of being likely true theoretically, but a weak claim in the sense of not being specific about directions for future research.

Therefore, I will offer some brief remarks about the feasibility of acquiring and representing encyclopedic knowledge at the linguistic level, and suggest some directions and possibilities for future research, while avoiding speculation.

Section 3.6.1 notes that WordNet (Miller, 1991) would be a natural resource to leverage in building a Tala lexicon. Likewise, Wikipedia would be a natural resource to leverage for the propositional knowledge typically found in human encyclopedias, and indeed, it has already been the subject of research on semantic analysis. For instance, Flickinger *et al.* (2010) discuss research on syntactic parsing and semantic annotation of Wikipedia. It appears in principle such an approach could create Tala sentences expressing encyclopedic knowledge from Wikipedia, though this remains to be explored in future research.

Lenat (1995) describes the Cyc project to represent commonsense knowledge, which began in 1984 and still continues. The effort has shown that such knowledge is vast, and time-consuming to acquire and represent, at least in the formal language (CycL) developed for the project. Sowa (2010) attributed Cyc's failure to achieve human-level artificial intelligence to the lack of a close connection with natural language, and to lacking a child's flexibility in learning and adapting information.

Since commonsense knowledge is outside the scope of this thesis, it will not comment on the extent to which Cyc may be useful for future research in the TalaMind approach. Here it will only be suggested that the TalaMind approach may avoid some of the problems Sowa described: Being based on natural language, Tala's support for underspecification, negation, paraphrase, metaphor, metonymy and idioms may enable representing commonsense knowledge more flexibly than is possible in existing formal languages (cf. §4.2.2.3).

Similar remarks apply to sociocultural knowledge. Being based on natural language, Tala can be used to describe interactive norms and goals in typical situations, enabling this form of encyclopedic knowledge to be represented in semantic domains (§3.6.7.7) at the linguistic level, and augmented by representation in semantic frames at the archetype level of Figure 1-1.

To a limited extent, Tala can also be used to represent perceptual and behavioral knowledge at the linguistic level. For instance, Tala executable concepts can describe how to perform procedures and describe where an agent should look for something, or what something looks like, etc. However, the lower levels of Figure 1-1 are needed to fully represent and implement perceptual and behavioral knowledge within a physical environment.

The prototype TalaMind demonstration system illustrates how a very limited amount of encyclopedic knowledge can support higher-level mentalities of human intelligence. Small semantic domains for encyclopedic knowledge about nuts, grain, and people are used to illustrate learning by reasoning analogically, causal and purposive reasoning, etc. (viz. Chapters 5 and 6).

3.6.7.5   Interactive Contexts and Mutual Knowledge

Extending the idea of dialog contexts, a Tala agent's conceptual framework should support representation and conceptual processing of interactive contexts, i.e. contexts for interaction of a Tala agent with humans and other Tala agents. The temporal dimension of an interactive context includes sequences of Tala sentences, communicated between Tala agents and humans. For example, playing a game, negotiating a business deal, working in a team, trying a court case, waging a political campaign, writing and reading a book are interactive contexts. All involve communication and mutual knowledge representation.

Bunt (2000) categorizes 'dialog acts' necessary for successful participation in a dialog. Besides task-oriented dialog acts (speech acts communicating information related to the topic of the dialog, the speaker's intentions and beliefs, etc.) there are 'dialog control acts' of three kinds:

- Feedback acts that allow participants to indicate whether they comprehend each other, or believe the other comprehends them.

- Interaction management acts that allow a participant to retract or correct remarks, pause, exchange turns, request confirmation, change topics, etc.

- Social obligations management acts including greetings, introductions, apologies, and thanks.

These have not been studied in this thesis but in principle should be possible to support, e.g. in a semantic domain giving Tala sentences as definitions for various dialog control acts. This is a topic for future research.

Bunt (2000) discusses the importance of representing mutual knowledge of beliefs and intentions, for natural language understanding in dialogs, noting that agents must have models of other agents' models of dialog context (including beliefs and intentions), recursively, to support dialogs.

A Tala agent's conceptual framework can support recursively nested conceptual contexts.[57] For example, a Tala agent can have an interactive context representing its belief and intentions within a dialog, and within this context have a nested context modeling another participant's beliefs and intentions, and so on. The mechanics for this also supports hypothetical contexts and nested conceptual simulation, discussed in

---

[57] Cf. Johnson-Laird (1983, p.433) re 'recursive embedding of mental models'.

the next section. A demonstration of this is given in Chapter 6.

Also, Tala sentences can provide a "lightweight" representation of mutual knowledge, allowing statements about mutual knowledge to be expressed declaratively, and used for inference. For example:

```
Peter knows Da Vinci painted the picture and knows John knows
it.
(know
      (wusage verb)
      (subj Peter)
      (obj
            (and
                  (paint
                        (<- ?p)
                        (wusage verb)
                        (tense past)
                        (subj ("Da Vinci" (wusage noun)
                                    (naming proper)))
                        (obj
                              (picture
                                    (wusage noun)
                                    (det the)
                                    )))
                  (know
                        (subj John)
                        (obj ?p)]
```

The `(<- ?p)` expression provides a coreference pointer, allowing the sentence to express that Peter knows Da Vinci painted the picture and knows John knows the same thing. Of course, this does not provide a complete model of John's knowledge, nor guarantee that Peter's knowledge about John's knowledge is correct. Also, the above expression does not describe infinite recursion of shared knowledge. That can be represented (finitely) as follows:

```
Peter knows Da Vinci painted the picture and knows John knows
it and knows John knows Peter knows it, ad infinitum.
(know
      (<- ?k)
      (wusage verb)
      (subj Peter)
      (obj
            (and
                  (paint
                        (<- ?p)
                        (wusage verb)
                        (tense past)
                        (subj ("Da Vinci" (wusage noun)
                                    (naming proper)))
                        (obj
                              (picture
                                    (wusage noun)
                                    (det the)
                                    )))
```

111

```
(know
        (subj John)
        (obj
                (and ?p ?k)]
```

It can also be important to declaratively represent that knowledge and beliefs are not shared (Silverman & Whitney, 2012):

```
Peter thinks Da Vinci probably painted the picture and thinks
the seller does not think so.
(think
        (wusage verb)
        (subj Peter)
        (obj
                (and
                        (paint (wusage verb)
                                (<- ?p)
                                (tense past)
                                (adv probably)
                                (subj ("Da Vinci" (wusage noun)
                                        (naming proper)))
                                (obj
                                        (picture
                                                (wusage noun)
                                                (det the)
                                                )))
                        (think
                                (wusage verb)
                                (adv not)
                                (subj
                                        (seller
                                                (wusage noun)
                                                (det the)
                                                ))
                                (obj ?p)]
```

3.6.7.6 Hypothetical Contexts

To support reasoning about potential future events, and counterfactual reasoning about past and present events, a Tala agent's conceptual framework should support creation and conceptual processing of hypothetical scenarios of events. Logically, these should have the same properties as event memory or perceived reality, and interactive contexts, with the difference that a Tala agent's conceptual processing can simulate the evolution of events within hypothetical contexts. Thus a hypothetical context may include models of other agents' beliefs and goals, to support simulating what they would think and do, hypothetically. The term *nested conceptual simulation* is used to refer to an agent's conceptual processing of hypothetical scenarios, with possible branching of scenarios based on alternative events, such as choices of simulated agents within scenarios. There may be no difference in the conceptual framework between mental spaces (§3.6.7.8) and hypothetical contexts, or a mental space may be a lightweight hypothetical context. This is a topic for design. Chapters 5 and 6 discuss nested conceptual simulation in the prototype.

3.6.7.7 Semantic Domains

Section 3.6.1 discussed the Tala lexicon for storing definitions of words within the conceptual framework, noting that the possible meanings of a commonly used English word are not bounded by a finite list of precise definitions (§2.2.2) and that a lexicon must be integrated with encyclopedic knowledge (§3.2.2).

This thesis will refer to "the encyclopedic knowledge to which a meaning of a word is linked" as a *semantic domain* for a sense of the word. The semantic domain includes the definition of the sense of the word, if there is a definition (viz. §3.6.8) and Tala sentences using that sense of the word, relating it to other concepts. Since a semantic domain is a collection of sentences, it is a type of context.

This description of semantic domains corresponds to Langacker's (1987 *et seq.*) description of conceptual domains in cognitive grammar, which he referred to simply as domains. Essentially the same idea of conceptual domains supports Lakoff and Johnson's (1980, 1999) theory of conceptual metaphors and metonyms.

This thesis uses the term 'semantic domain' rather than 'conceptual domain', to specifically include Tala sentences in semantic domains at the linguistic level of Figure 1-1, stipulating that Tala sentences can refer to concepts in the archetype, level. The next two sections will discuss semantic domains in support of mental spaces, conceptual blends, and conceptual metaphors.

Multiple words can each provide access to a common semantic domain, and a word may have more than one sense in a semantic domain, e.g. if used as a noun and a verb. The concepts in a semantic domain can often be defined or described using word senses that occur in other semantic domains. The set of semantic domains thus referenced by a semantic domain is called its 'domain matrix' (cf. Langacker, 1987).

Evans (2009) notes that linguistic context helps limit the encyclopedic knowledge relevant to a word's usage. This implies a theoretical requirement that a semantic domain can be accessed by specifying a combination of words or Tala expressions provided by linguistic context (interactive contexts, perceived reality, etc.). It suggests the potential for TalaMind semantic domains to support domain-driven word sense disambiguation (Gliozzo *et al.*, 2004; Buitelaar *et al.*, 2006), though this is a topic for future research. In this regard, a TalaMind semantic domain is compatible with the description given by Gliozzo and Strapparava (2009, p.6), i.e. words belonging to the same semantic domain denote concepts that are strongly related to each other.

Fillmore (1977, 2002) uses the term 'semantic domain' with compatible meaning referring to the subject areas of 'semantic frames', his theoretical description of conceptual structures needed to understand meanings of a group of related words (Fillmore, 1975 *et seq.*). Fillmore's semantic frames are complementary to Langacker's domains, with additional focus on concepts for interactive norms and goals in situations or scenes – conceptual content also discussed by Minsky (1974) and Schank and Abelson (1977). Per §3.6.7.4, encyclopedic knowledge includes sociocultural knowledge about interactive norms and goals. Hence, such concepts also exist as Tala sentences in TalaMind semantic domains, at the linguistic level of Figure 1-1.

Since encyclopedic knowledge can have various epistemic modes (§3.6.7.1), different semantic domains can have different epistemic modes. For instance, the word "klingon" is linked to a fictional semantic domain, while the word "human" is linked to a factual domain (viz. §4.2.8).

Since a TalaMind semantic domain is a context, in principle inference can be performed within it, i.e. using the sentences already in the semantic domain as premises to derive new sentences that can be added to it. To the extent that semantic domains and encyclopedic knowledge are global (i.e., relatively constant as described in §3.6.7.1) such inference will be relatively infrequent, compared to inference in local contexts.

3.6.7.8 Mental Spaces

Of course, constructing interpretations for sentences involves much more than retrieving definitions of words and encyclopedic knowledge in semantic domains. A human-level AI must construct meanings for sentences that are appropriate relative to its perceived reality, event memory, and ongoing interactive contexts. While inference (§3.6.6) is clearly required for disambiguation and meaning construction, it is desirable to identify and support other specific conceptual processes for constructing meaning of natural language sentences, especially if such processes can also support other higher-level mentalities. This motivates Hypothesis III's inclusion of mental spaces and conceptual blends in the TalaMind approach. The present section discusses mental spaces as a context type in TalaMind architectures, while the next discusses conceptual blends. Their usage to support higher-level mentalities is discussed in §3.7.

Evans and Green (2006, pp.363-396) outline Fauconnier's (1984 *et seq.*) theory of mental spaces. Unlike semantic domains, which are global, relatively static contexts of encyclopedic information, mental spaces are dynamic, local, temporary contexts created by a Tala agent when understanding natural language sentences. Constructing an interpretation for a single natural language sentence can cause the creation of multiple mental space contexts. These can be reused and extended to understand subsequent natural language sentences that occur in an interactive context (§3.6.7.5). An agent can import information from a semantic domain into a mental space and perform inference within a mental space.

At the linguistic level in a TalaMind architecture, mental spaces are collections of Tala sentences, and therefore a context type. Additionally, a mental space identifies a set of 'elements', which are entities represented by noun expressions or pronouns. These elements are either constructed while processing a discourse, or pre-existing in the conceptual framework. For instance, elements may correspond to noun expressions described by encyclopedic knowledge. The Tala sentences within a mental space express relationships between its elements.

Mental spaces theory describes how mental spaces are created during conceptual processing of a natural language sentence. Linguistic expressions called 'space builders' can invoke creating a mental space, or move attention between mental spaces. For example, connectives (*if X then...*), prepositional phrases (*in theory, in 1949, at the park*), and adverbs (*maybe, certainly*) can be space builders. Subject-verb pairs (*John thinks...*) and nested sentences (*Pat says [Jane enjoys opera]*) can also be space builders (viz. Fauconnier, 1994, pp.16-18; Evans & Green, 2006).

Mental spaces theory says that mapping relationships (also called connectors) between the elements of different mental spaces are created to help represent interpretations of a natural language sentence. These mappings can be represented by Tala sentences.

Considering the above descriptions as requirements, a mental space can be represented minimally within a TalaMind architecture by a data structure such as:

```
(<space-number> ;a unique identifier for each mental space
    (elements ;Tala noun concepts
    )
    (concepts ;Tala sentences
    ))
```

A set of connectors between mental spaces can be represented minimally by a list of Tala sentences. ('Minimally' means that additional information could be included in these structures, or other data structures could be used if needed, e.g. for scalability.)

For example, Fauconnier (1985, 1994; pp.36-37) discussed how mental spaces could represent a situation in which Orson Welles portrays Alfred Hitchcock in a biographical movie about Hitchcock, while a minor role in the movie (a man at a bus stop) is played by Hitchcock personally.

Following are TalaMind data structures representing two mental spaces and associated connectors for this situation. The first mental space is created for the initial discourse context of the speaker's reality, the second represents the imaginary movie.

```
(1
    (<- ?s1)
    (relative-time present)(epistemic-mode reality)
    (elements
        ("Orson Welles"
            (<- ?a1)
            (wusage noun)
            (naming proper)
            )
        ("Alfred Hitchcock"
            (<- ?b1)
            (wusage noun)
            (naming proper)
            )
        )
    (concepts
        (suppose (wusage verb)
            (obj
                (make (wusage verb) (tense present)
                    (passive)
                    (subj ?m)
                    )))))
(2
    (<- ?s2)
    (relative-time
        (before ?s1))
    (epistemic-mode imaginary)
    (elements
```

```
("Alfred Hitchcock"
      (<- ?b2)
      (wusage noun)
      (naming proper)
      (in ?m)
      )
 (man
      (<- ?b3)
      (wusage noun)
      (det the)
      (in ?m)
      (at
            (stop (wusage noun)
                  (comp (bus (wusage noun)))
                  (det the)
                  )))
 (movie (wusage noun)
      (<- ?m)
      (det a)
 (about ?b1)
      )
      )
(concepts
    ;Hitchcock shoots the man at the bus stop.
    (shoot (wusage verb)
          (subj ?b2)
          (obj ?b3)
          (in ?m)
          )
    ))

(connectors
    ;Orson Welles plays Alfred Hitchcock in the movie.
    (play (wusage verb)
          (wsense acting)
          (subj ?a1)
          (obj ?b2))
    ;Alfred Hitchcock plays the man at the bus stop
     ;in the movie.
    (play (wusage verb)
          (wsense acting)
          (subj ?b1)
          (obj ?b3))
    ;Hitchcock in the movie portrays Hitchcock
     ;in reality.
    (portray
          (wusage verb)
          (subj ?b2)
          (obj ?b1))
    ;The mental space ?s2 represents the movie
    (represent
          (wusage verb)
          (subj ?s2) (obj ?m)]
```

Fauconnier uses this example (presented as a diagram) to show that different elements representing Hitchcock must be provided in the mental spaces to represent unambiguous interpretations of sentences like "The man at the bus stop is shot by Hitchcock, in the movie".

In the above examples, two attribute slots are added to each mental space to describe their relative time and epistemic modes. Evans and Green (2006, pp.387-396) discuss in some detail the research of Cutrer (1994) and others regarding mental spaces and the English tense-aspect-modality system.

There is much more that can be said about mental spaces, and which has been said by Fauconnier and others, regarding their use in meaning construction for natural language. However, the above paragraphs will suffice to support discussions throughout this thesis. There do not appear to be any theoretical issues that prevent representation and use of mental spaces within a TalaMind architecture.

3.6.7.9 Conceptual Blends

While inference is clearly required for meaning construction, it is desirable to identify and support other specific conceptual processes for understanding natural language, especially if such processes can also support other higher-level mentalities. And as discussed in §3.6.6.5, interpreting meaning for metaphors is essential for understanding natural language as well as humans do. Conceptual blends are an extension of mental space theory developed to explain how metaphors are understood, which are claimed to support human imagination (Fauconnier & Turner, 1994 *et seq.*). These considerations motivate Hypothesis III's inclusion of conceptual blends in the TalaMind approach. This section describes conceptual blends, focusing on how they support metaphors. The use of blends to support higher-level mentalities is discussed in §3.7.

Evans and Green (2006, pp.400-440) outline Fauconnier and Turner's theory of conceptual blends, saying that international academic research supports a view that conceptual blends are important for imagination and thought in the human mind.

Fauconnier and Turner (2002, p.46) describe a conceptual blend 'integration network' including four mental spaces: a generic space of background information, two input spaces, and a space that integrates (blends) concepts from the two input spaces. There are connectors between the spaces specifying relationships between elements of the spaces. The integration network also includes connectors between the input spaces, which specify relationships used by conceptual processing to construct concepts in the blend space.

The generic space subsumes concepts available to interpret a metaphorical expression: encyclopedic and commonsense knowledge, perceived reality, event memory, ongoing interactive contexts, etc. The input spaces contain concepts retrieved from semantic domains identified in the natural language expression being interpreted. For the 'surgeon as butcher' metaphor there is an input space for 'surgeon' and another for 'butcher'. Conceptual integration networks can have more than two input spaces if needed.

Blending theory describes a series of conceptual processes that construct integration networks and integrate concepts within them:

- A *matching* process performs a partial matching to identify counterpart elements in the input spaces and establish cross-space connectors between them. These counterpart connectors can be based on several kinds of 'vital relations' between the elements, e.g. identity, role, cause-effect, etc., as well as metaphoric specification.

- *Selective projection* is a conceptual process that projects concepts from the input spaces into the blend space. Not all the concepts in the inputs are projected to the blend, only those related to the metaphorical expression being interpreted, and to the cross-space connectors between the input spaces.

- *Composition* is a conceptual process that composes separate elements from the input spaces into a single element in the blend space. For the 'surgeon as butcher' blend this process creates concepts representing a surgeon who has the goals, skills, and tools of a butcher, in the blend space.

- *Completion* is a conceptual process that recruits additional concepts from background information into the blend space, prompted for by composition or the next process, elaboration. For example, in the 'surgeon as butcher' blend, completion might recruit additional encyclopedic and commonsense knowledge into the blend, to support a conclusion that patients would die in operations performed by a surgeon with the goals, skills, and tools of a butcher.

- *Elaboration* is a conceptual process that develops additional concepts in a blend using imaginative simulation (Fauconnier & Turner, 2002, p.48). This involves inference within the blend, and is the same kind of process as nested conceptual simulation in hypothetical contexts (§3.6.7.6). For the 'surgeon as butcher' blend, this process would conclude that a surgeon with the goals, skills, and tools of a butcher would be an incompetent surgeon (Grady *et al.*, 1999).

Researchers have studied governing principles and constraints for these processes, which are not deterministic: Different blends can be created from the same input spaces by different speakers or by the same speaker at different times (Evans & Green, 2006, p.409).

Together, these conceptual processes produce emergent conceptual structures in the blend, concepts not specified in the original metaphorical expression. Other contexts can be modified as a result. Concepts from the blend can be projected backwards to the input spaces or to contexts in the generic space. For the 'surgeon as butcher' metaphor, a disanalogy relationship between the input spaces can be created, and the original interactive context can be modified, attributing to its speaker a negative comment on the surgeon's competence. Though the metaphor was understood dynamically, if it is salient for other situations it can become conventionalized so that future interpretation is automatic, e.g. using a construction to perform metaphoric translations (§5.4.13).

Conceptual blends are complementary to Lakoff and Johnson's (1980, 1999) conceptual metaphor theory in several respects (Grady *et al.*, 1999; Evans & Green, 2006). This theory discussed how metaphors can be represented by unidirectional mappings between conceptual domains, stored in long-term memory, but did not discuss dynamic, temporary mental spaces or blend spaces. In principle in a TalaMind architecture, conceptual metaphors can initially be understood using mental spaces with concepts projected from semantic domains. If salient, metaphors can become conventionalized and used for relatively stable knowledge representation in semantic domains. Vogel (2011) gives a formal logic approach to metaphors.

Lakoff and Johnson (1980, 1999) also presented a conceptual theory of metonymy, noting that a metonymical reference allows one entity to stand for (provide access to) another in the same conceptual domain. This is a different kind of mapping from that involved in metaphor (Evans & Green, 2006, p.312; Kövecses & Radden, 1998). It can also be supported in a TalaMind architecture, in principle: Since metonyms are *ad hoc*, they can be disambiguated within a temporary mental space, using concepts projected from a semantic domain, i.e. metonyms do not require a four-space conceptual integration network.

There does not appear to be any *a priori* limit on the nature or form of metaphors that can be interpreted using conceptual integration networks. For example, XYZ metaphors ("Variety is the spice of life", "A metaphor is a bridge between worlds"[58]) can be interpreted using conceptual blends. Besides metaphors, conceptual integration networks can be used for other aspects of natural language understanding, e.g. counterfactual semantics, compound words, etc. (viz. Evans & Green, 2006).

There is much more that can be said about conceptual blends, and which has been said by Fauconnier and Turner, and others, regarding meaning construction for natural language. However, the above paragraphs will suffice to support discussions throughout this thesis. There do not appear to be any theoretical issues that prevent representation and use of conceptual blends within a TalaMind architecture, using its representation for mental spaces (§3.6.7.8). Use of conceptual blends in the TalaMind approach does not preclude other methods for interpreting metaphors.

3.6.7.10 Theory Contexts

A human-level AI must have the ability to develop theories, and to reason with and about theories, so that it can create theoretical explanations and predictions of its environment, to behave intelligently within its environment. Hence it must have the ability to represent theories.

An explanation or theory can sometimes be stated in a single sentence, e.g. "Everything is made out of earth, air, fire, and water." Yet we can develop more complex theories that support multiple conclusions and are reusable in multiple situations. Such theories can be stated as collections of Tala sentences, and are therefore a type of context, which this thesis calls (TalaMind) theory contexts, or simply 'theories'.

A TalaMind theory context includes the following kinds of

---

[58] One could also say "A metaphor is a figure of speech", as a recursive XYZ metaphor, though 'figure of speech' is engrained as a linguistic term and its own metaphorical nature may be overlooked.

sentences:

- An initial set of Tala sentences giving definitions of terms (word senses) used in the theory. A theory may also use 'undefined' terms that do not have definitions included in the theory. Undefined terms may or may not have definitions outside the theory.

- An initial set of Tala sentences stating relationships between the terms of the theory. This set of sentences can have various names, such as axioms, premises, or hypotheses – in this discussion the name does not matter, and the term 'premises' will be used.

- A set of Tala sentences that can be derived directly or indirectly from the definitions and premises of the theory. This set of sentences also has various names, such as theorems, predictions, conclusions, etc., which do not matter in this discussion.

- A set of Tala sentences expressing conjectures, questions, and derivations for sentences in the theory. A derivation can be expressed using the Tala syntax for a sequence of steps, which is also used in an executable concept. Thus, a derivation can be represented as a conceptual structure, available for conceptual processing to verify or dispute.

Conceptual processing is responsible for deriving conclusions in a theory context from the premises, definitions, and previous conclusions derived in the context. Since Tala sentences are based on natural language, a TalaMind theory context has the potential for ambiguity and underspecification found in natural language. So, it is not limited to the semantics of a formal theory stated in predicate calculus, or other precise logical formalisms. TalaMind theory contexts provide a general representation of informal, natural language–based theories. Per §3.2.1, this does not preclude the use of formal, logical languages in mathematical or scientific theories.

When it is initially developed, a theory context is a 'local', temporary context. If it is successful as a theory, it will eventually become a 'global' context for a Tala agent, included in its encyclopedic knowledge. Depending on the nature of the concepts involved, it may or may not be possible for a Tala agent to augment or express a TalaMind theory context by a formal theory, in a precise mathematical or logical language. As will be discussed in §3.6.7.14, a theory context can also have a meta-context for meta-reasoning about the theory. The TalaMind approach permits a meta-theory context that is about multiple other theories.

3.6.7.11 Problem Contexts

Within a context a Tala agent may have goals, which can be represented as Tala sentences of the form "I want X", where X does not yet exist in the context. Achieving a goal may be considered a problem, and the context considered a 'problem context'. If there is not yet a known way to achieve a goal, then an agent can reason about ways to achieve it. This may involve nested conceptual simulation within the problem context. If a solution is discovered, it may be described as a Tala sentence, for reuse in future similar problem contexts. This sentence will typically be of the form "If I want X then do Y" where Y may be a verb expression, and the verb may be defined by an executable concept.

Meta-statements about how to find solutions in problem contexts are of interest theoretically. McCarthy *et al.* (1955) termed such statements 'rules of conjecture' and in 1980 McCarthy proposed circumscription as a rule of conjecture to address the qualification problem in representing commonsense knowledge. Circumscription is equivalent to a meta-statement saying a problem context contains all the facts relevant to solving a problem, and that no new entities or facts may be introduced into the context to determine how to solve it. This is a relevant meta-statement when one wishes to consider whether a problem is logically solvable or unsolvable, and, if it is solvable, what is the most efficient solution.

In real-world problem contexts one may have a different rule of conjecture, for example that the problem context includes the initial facts known to be relevant to stating the problem, but not necessarily all the facts relevant to a solution. This is a more accurate statement for practical and scientific problems when one wants to find any possible way to solve a problem and then evaluate its practicality. So, if a real-world problem is isomorphic to the Missionaries and Cannibals problem, it may be completely legitimate to consider whether a bridge exists nearby.

Just as Tala is useful for semantic inference with underspecification, it may be useful for problem solving without considering all possible conditions for performing an action. Section 4.1.1 discusses this topic further, in addressing Dreyfus' criticism of AI regarding the qualification problem. Chapter 6 illustrates how the TalaMind approach can support problem-solving in the demonstration system.

3.6.7.12 Composite Contexts

In a TalaMind architecture, contexts will need to be constructed dynamically that combine aspects of the context types described above. For instance, a context may need to be a hypothetical interactive problem context, based on perceived reality and event memory. It is theoretically possible to have composite contexts if attributes of context types are consistent (viz. §3.6.7.1).

3.6.7.13 Society of Mind Thought Context

Per §§1.5 and 3.2.2, the conceptual framework has an implied requirement to support storing and retrieving thoughts represented as sentences in Tala. Any Tala sentence may be considered as a thought when it is created in any context within the conceptual framework. However, if a TalaMind system includes a society of mind architecture (§2.3.3.2) in which subagents communicate using the Tala conceptual language, then it is implied that these subagents will have an interactive context (§3.6.7.5). This will be termed a society of mind 'thought context'. The TalaMind hypotheses do not require a society of mind architecture but it is consistent with the hypotheses and natural to implement one at the linguistic level of Figure 1-1. Hence this chapter does not present theoretical requirements or analysis related to a society of mind. The prototype TalaMind architecture presented in Chapters 5 and 6 includes a simple society of mind in which multiple subagents communicate in a thought context.

3.6.7.14 Meta-Contexts

McCarthy (1995) noted the importance of being able to transcend a context and think about a context. Accordingly, this thesis stipulates that in addition to being able to perform inference within a context, using the sentences already in the context, a human-level AI needs to be able to perform inference about a context. Logically, the collection of sentences expressing statements about a context A is context about the context A, i.e. a meta-context for A. Thus the TalaMind approach stipulates that meta-contexts are needed. This implies, incidentally, that it must be possible to refer to a context by a name or a pointer, which is not theoretically problematic in a TalaMind architecture. The example given in §3.6.7.8 uses the (<- ?p) pointer notation for references to mental space contexts.

The attributes and features of a context can be expressed by Tala sentences in its meta-context. For example, if a Tala agent determines that a context contains a contradiction, it can create a sentence saying that the context contains contradictory sentences. Similarly, a Tala agent can have sentences in a meta-context expressing rules of conjecture for a problem context (§3.6.7.11), or describing the epistemic mode of a semantic domain (§§3.6.7.7, 4.2.8).

The question arises, how to avoid an infinite regress of contexts about contexts? The TalaMind approach does not require or imply any particular method for this. One approach would be to not have any preset limit, but to effectively limit the levels of meta-contexts by a resource constraint making it increasingly less likely an agent will create higher levels of meta-contexts. Weyhrauch (1980) proposed another approach, in which a reserved context can support an indefinite amount of meta-reasoning about other contexts.

A third approach is to not create a meta-context outside a context, but to perform meta-reasoning about a context within the context itself. In principle, a Tala expression can refer to the context in which it occurs: If the (<- ?p) pointer notation is used to bind a pointer ?p to a context, then Tala expressions within the context can be statements about the context, using ?p as a referent. This approach would enable a natural language of thought to support metacognition within a context. By analogy, people often say "I like/don't like this situation" to make a meta-statement about the context they are in.

The nature and representation of meta-contexts are topics for future research.

3.6.8 Primitive Words and Variables in Tala

? Does Tala contain primitive words, and if so, how are their meanings represented and determined?

This question has confronted previous work on formal concept representation languages for AI. For example, Dreyfus (1992) noted that Schank's work on conceptual dependency and scripts seemed to face an endless task – initial primitives might be specified, but if they were not sufficient to represent human concepts then more primitives might need to be added.

This problem does not confront the TalaMind approach in quite the same way it did for Schank. For Tala to support understanding natural language, we do not need to stipulate ad hoc, context-free primitives in a formal language for representing the semantics of words separate from natural language. However, we do need to address the topic of primitives in natural language itself, i.e. primitive words in a particular natural language such as English. This question will be given a case-by-case answer, for different grammatical categories of English words.

Since new words are typically defined in terms of previous words, it is natural to think there is some set of words that are most basic, or primitive, in terms of which other words are ultimately defined. If people try to define such words, they may create circular definitions, or define them by reference to context, e.g. pointing to objects or events in the environment. Other primitive words, e.g. prepositions, conjunctions, and articles, are effectively built into the syntax of English, and may be even more difficult to define.

It is possible to create a list of proposed primitive words for English (or any other natural language) and several researchers have studied this topic. There is substantial agreement in primitives suggested by several sources, including Bateman *et al.*, Dolan *et al.*, Jackendoff, Kay and Samuels, Mandler, Masterman, Wierzbicka and Goddard, and Zholkovsky. Benedetti (2011, pp.17-18) reports comparing Wierzbicka's primitives with the Swadesh words used in glottochronology to determine kindred relationships of languages, writing that the Swadesh list is used in almost all languages. Benedetti finds substantial agreement in both sets of words, though they were developed for different purposes.

However, different words may be considered primitive, from different theoretical perspectives. For example, the Swadesh list of basic words includes man, woman, bird, dog, etc. It does not include electron or atom, which from a physics perspective could be considered more primitive. Nor does Swadesh contain abstract terms such as entity or relation, which could be considered primitive for an ontology.

In a TalaMind architecture, all nouns are equally accessible, and reasoning can occur with each of them directly, without necessarily having to reason about how a noun is defined in terms of other words. A noun may be used as an index for a semantic domain, which will provide encyclopedic and commonsense knowledge about it. This information will be available for abductive reasoning to disambiguate its sense and reference, supported by lower levels of conceptual processing and the environment interaction systems (Figure 1-1). Disambiguation of a noun is a form of variable binding, as discussed in §3.6.1. So, the TalaMind approach does not require or depend on a list of primitive nouns for English, *per se*, although if the senses of Tala nouns are defined in semantic domains that are organized in a lattice, and circular definitions are not allowed, then it follows that some nouns will not be defined in terms of other nouns.

The semantics for articles (a, an, all, some, most, the, …) and conjunctions (or, and, if, then, while, …) is determined by the conceptual processing of Tala sentences that contain these words. That is, there is no need to define these primitive words in terms of other words. Their meaning derives from how they are interpreted and/or semantically annotated during conceptual processing. This will often involve underspecification, per §3.6.3.7.

The semantics of pronouns depends on abductive disambiguation performed by conceptual processing, to determine most likely references in context. That is, these primitives are disambiguated as context-sensitive references, which may be represented using pointers either within Tala concept structures, or to percepts in the conceptual framework. To illustrate, here is a Tala interpretation of an example from Bunt and Black (2000, p.9):

```
(if
    (thrive
        (wusage verb)
        (adv not)
        (subj
            (baby (wusage noun)(det the)))
        (on
            (milk
                (<- ?m)
                (wusage noun)
                (from
                    (cow (wusage noun)(number plural)))
                )))
    (then
        (boil
            (wusage verb)
            (obj
                (it (wreferent ?m)]
```

The semantics of prepositions is more complex but also depends on abductive disambiguation by conceptual processing, to disambiguate in context. Two classes of prepositions may be distinguished: those that appear to have a collection of somewhat distinct, different meanings; and those that have broadly varying meanings, and serve mainly to guide interpretation in context. In the first class are prepositions like 'over', for which Tyler and Evans (2003) found fifteen distinct meanings – abductive disambiguation with encyclopedic knowledge is needed to select the most likely meaning in context. The preposition 'of' is in the second class. It indicates a variable relationship between two entities. The nature of the relationship depends on the specific entities and must be determined by abductive conceptual processing in context.

For verbs the situation is often the same as for nouns in the TalaMind approach. That is, verbs can also be indexes for semantic domains, and reasoning about the meaning of a verb can occur in much the same way as for a noun. However, certain primitive verbs such as 'is', 'have', 'do', 'go', etc. are exceptions to this. These can have variable meanings requiring abductive disambiguation in context, similar to the disambiguation of prepositions described above. This will be illustrated in Chapters 5 and 6.

The TalaMind approach is similar for adjectives and adverbs: disambiguation involves abductive reasoning with encyclopedic knowledge in context, to select the most likely meanings, resulting in translation of conceptual structures (cf. §3.6.3.9).

Tala will need to have some primitive, undefined reserved words and variables not specifically related to the semantics of English, e.g. to support definition of executable concepts. Chapter 5 discusses some initial primitives used in the demonstration system, such as the reserved variable `?self`, function names used as primitive verbs to invoke constructions, etc.

3.7 Higher-Level Mentalities

The preceding sections have discussed how the TalaMind approach could support natural language understanding, which is one of the higher-level mentalities listed in §2.1.2. This section considers the following theoretical questions for the other higher-level mentalities:

?   What is theoretically required, to claim a system achieves each higher-level mentality?

?   How can each higher-level mentality be supported by the TalaMind hypotheses and architecture, at least in principle theoretically?

3.7.1 Multi-Level Reasoning

Following are theoretical arguments that the TalaMind architecture can perform each of the kinds of reasoning included within multi-level reasoning as defined in §2.1.2.6.

3.7.1.1 Deduction

Deduction has classically been described using natural language expressions, e.g.:

> All men are mortal.
> Socrates is a man.
> Therefore, Socrates is mortal.

As noted in §3.6.6, Tala's use of nested list structures to represent natural language syntactic structures facilitates pattern-matching of Tala sentences. Combined with a syntax for variables in Tala, and representation of inference rules as if-then sentences, these symbolic processing mechanisms enable logical deduction, within contexts.

3.7.1.2 Induction

If a Tala agent develops a set of observations that support the truth table corresponding to $A \rightarrow B$, then the agent may conditionally, though not certainly, infer the rule $A \rightarrow B$ by induction.

The introduction of conditional inference and associative observations opens the door to probabilistic inference (Pearl, 1988 *et seq.*). Moreover, Pearl suggests Bayesian logic should be integral with the representation of contexts, that according to Bayesian philosophy the condition $A \mid B$ means A in a context indicated by B (viz. Pearl, 2009, p.4). This would be an interesting topic for future research in the TalaMind approach.

3.7.1.3 Abduction

Abduction is a reasoning process that develops explanations (see §2.2.2):

> C is true
> $A \rightarrow C$
> Therefore, perhaps A is true.

Abduction is possible within a TalaMind architecture, using the same symbolic processing mechanisms that support deduction. When abducing A as an explanation for C, where does the rule $A \rightarrow C$ come from? One answer is to construct rules via inductive observation, as described above; another is to use analogy to construct new rules – see next.

3.7.1.4 Analogical Reasoning

Analogical reasoning is of the form:

> $A \rightarrow B$
> C is similar to A
> Therefore perhaps $C \rightarrow B$

Or more generally:

> $A \rightarrow B$
> C is similar to A
> D is similar to B
> Therefore perhaps $C \rightarrow D$

This is also called case-based reasoning, if $A \rightarrow B$ is a previous case, and we are trying to determine how to process C. Similarity may be recognized at the archetype or associative levels in Figure 1-1, or it may be determined based on linguistic descriptions, e.g. a system may consider that "date wine" is similar to "grape wine", since both are linguistically described as types of wine (cf. Sowa & Majumdar, 2003). Examples of analogical reasoning within a TalaMind architecture will be given in Chapter 6, for the prototype demonstration system.

3.7.1.5 Causal and Purposive Reasoning

Natural language syntax enables concise, direct representation of causal and purposive relationships, e.g. via subordinating conjunctions ("because", "since", "why", "how", ...), or via verb or noun expressions ("X causes Y", "The purpose for doing X was Y"), etc. Thus Tala enables one to easily say "The O ring shattered because the temperature was below freezing", i.e. to state a causal relationship between arbitrarily complex expressions, with the causal link being accessible for pattern-matching and inference.[59]

Since these relationships can be directly expressed in natural language syntax, and in the Tala conceptual language, reasoning about

---

[59] As noted by Sowa & Majumdar (2003), causal and purposive reasoning may be used to determine whether analogical inferences are relevant.

such relationships can be supported within the TalaMind architecture. Examples of this will be given in Chapter 5 and 6, for the prototype demonstration system. The TalaMind approach is also open to use of probabilistic logic for reasoning about causality (Pearl, 2009) as a topic for future research.

3.7.1.6   Meta-Reasoning

Meta-reasoning is reasoning about reasoning, so it includes reasoning about reasoning processes, reasoning about the validity of arguments, deciding to abandon a theory, etc. In principle, it is theoretically possible to perform meta-reasoning within the TalaMind architecture, because:

- A concept represented by a Tala expression can refer to other concepts. A Tala sentence can declaratively express knowledge about knowledge (§3.6.7.5).

- Per §§3.6.7.10 and 3.6.7.14, TalaMind systems can in principle reason about theories, treating them as objects that have properties such as consistency or inconsistency.

- Executable concepts in outer contexts can access ("observe") concepts in nested contexts (§3.6.7.6), using observations for decision logic and meta-concepts.

- Reasoning by analogy may be combined with causal reasoning, as well as purposive reasoning about why something should be done, to determine if an activity is worthwhile.

- Conceptual processes can create concepts to record traces of their execution, which can be the subject of observation and reasoning by other conceptual processes.

Meta-reasoning supports recognition of paradoxes and termination of unproductive mental loops.

3.7.2   Self-Development and Higher-Level Learning

3.7.2.1   Learning by Multi-Level Reasoning

While induction is considered a form of inference, it is also a form of learning. Thus Holland, Holyoak, Nisbett, and Thagard (1986, p.1) consider induction to include expansion of knowledge in uncertainty. Processes for induction of new inference rules can be extended to learning new concepts in general. New concepts may be described in

Tala sentences at the linguistic conceptual level in the TalaMind architecture, or may be learned and represented at the archetype and associative levels.

Other forms of inference, such as abduction, analogical reasoning, and causal and purposive reasoning, enable corresponding forms of learning. Given the abductive ability to create an explanation, a Tala agent can make predictions based on such explanations. The success or failure of predictions can be used to improve explanatory theories. Likewise, reasoning by analogy enables developing explanations and theories for new domains based on analogies with previously known domains. Causal and purposive reasoning enables developing causal and purposive explanations, which again supports testing of predictions and improving explanatory theories. Thus, reasoning combined with experimentation can be used to support higher-level learning, leveraging the domain-independence and semantic generality of the Tala language. Chapters 5 and 6 illustrate this in the prototype demonstration system, showing how a Tala agent could learn about a new domain (making grain edible) via analogies with a previously known domain (eating nuts).

3.7.2.2   Learning by Reflection and Self-Programming

This form of higher-level learning (§2.1.2.5) includes:

- Reasoning about thoughts and experience to develop new methods for thinking and acting.

- Reasoning about ways to improve methods for thinking and acting.

Several capabilities can be identified as requirements for learning by reflection and self-programming in human-level AI: To reflect, a human-level AI must be able to recall memories of past events, and imagine how things could have happened differently, in a hypothetical world. For reflection to be effective (for it to "matter") the system must be able to examine its own behaviors (including its thought processes and speech acts), imagine how these might be improved, and then change its future behaviors – hence effective reflection involves self-programming. For reflection to be successful, the system must have some representation of goals relative to which it can evaluate success, and it must be able to reason about how its behaviors relate to achieving its goals. For reflection to be effective and successful at a level corresponding to human-level intelligence, the system must be able to represent and reason with any concept that can be expressed in natural language.[60] For reflection to be effective and successful, a human-level AI system must be able to attribute significance to events that it observes, and attribute meaning to what it hears and reads, and to what it does, says, and writes. It must use abductive reasoning to understand the meaning of others' words and actions, and causal reasoning to understand the consequences of its own thoughts and actions. Hence, effective reflection in human-level AI involves the ability for computers to have *semantic originality*, i.e. to be able to attribute meaning to words and actions, independently of human observation (viz. §2.3.5).

In principle, all of these can be supported within the TalaMind architecture. Recollection of past events is supported by an event memory, and counterfactual imagination is supported by nested conceptual simulation in hypothetical contexts. Recollection and analysis of previous thoughts could be supported by storing mental speech acts ('meffepts')[61] in event-memory, with trace information about the executable concepts that produced them. Goals can be represented as Tala sentences within contexts. The Tala conceptual language includes executable concepts that have the ability to create and modify executable concepts. Chapter 6 illustrates some of these theoretical capabilities in the prototype demonstration system, showing how a Tala agent could improve its method for making bread, by reasoning about how to change the process for making flat bread to create a new process that makes leavened bread.

The TalaMind architecture supports abductive and causal reasoning to understand the meaning and importance of words and actions. A Tala agent can declaratively represent that words have meanings, with definitions in Tala or with declarative statements about usage of a word in a semantic domain. Senses and referents of words can point to cognitive concept structures that in turn are grounded in perceptions of the environment (Figure 1-1). Likewise a Tala agent can declaratively

---

[60] These requirements imply the system should be able to refer to itself, or as Smith (1982) wrote, "we want the process as a whole to be able to refer to its whole self." In the TalaMind architecture the variable `?self` is a reserved name allowing a Tala agent to refer to itself.

[61] See the Glossary and §§5.4.3, 5.4.4 regarding speech acts and mental speech acts.

describe relations between its actions and their consequences. Thus a Tala agent can declaratively represent knowledge of what its words and actions mean. In principle it can develop this knowledge autonomously, independently of human observation or action, by virtue of the computational universality of TalaMind conceptual processes. Further, TalaMind conceptual processes can actively use and apply declarative knowledge about the meanings of words and the consequences of actions. Thus in principle a Tala agent can have semantic originality. Note that this does not mean a Tala agent must have knowledge and understanding of every detail of its construction and processing at lower levels of representation, just as semantic originality does not require humans to have such knowledge about themselves.

If executable concepts and conceptual processes are computationally universal, then the ability to develop new methods for human-level AI is very general, in principle. Reflection and self-programming can be guided within the TalaMind architecture by mechanisms for meta-reasoning and learning. Further, a Tala agent's reasoning about how to create new methods can leverage new languages and representations for a domain, addressing Minsky's knowledge representation problem for baby machines.

3.7.2.3   Learning by Invention of Languages

McCarthy (1955) noted that English can be used to define other languages, which can then be used as appropriate. Per §3.2.1, the Tala conceptual language retains this ability of English. Per Hypothesis III, Tala includes constructions to support extensions.

Language and representation have two important aspects for learning in the TalaMind approach. The first is that a conceptual language both constrains and facilitates expression of concepts, which has a corresponding effect on what concepts can be learned. The second is that a key to success in solving a problem, or learning about a domain, is often the ability to represent certain problem- or domain-specific concepts, i.e. to find and use a particular representation for the problem or domain.

Historically, one of the best examples of the first aspect was the co-invention by Newton and Leibniz of different mathematical languages for representing concepts in differential and integral calculus. The clarity and concision of Leibniz' notation was superior to Newton's, which had important consequences for their respective use in learning about new domains.

A classic example of the second aspect is the 'mutilated chessboard problem' (Gamow & Stern, 1958), which asks whether an NxN chessboard can be exactly covered by 2x1-sized dominoes, after removing two diagonally opposite corners of the board. It is exponentially hard to solve the problem by searching over all possible ways to tile the chessboard with dominoes. But if one can represent certain concepts it is easy to prove the problem is unsolvable. The concepts that need to be represented are: a) every domino must cover a black and white square; b) the mutilated chessboard will have unequal numbers of black and white squares.

Since Tala includes the ability to say "X means Y" to support extending English, and since any symbolic data structures and syntax could be defined as extensions, there is in principle no limit to the nature of new languages that could be supported. More specifically, the invention of new languages and representations can be guided within the TalaMind architecture by mechanisms for meta-reasoning and learning; that is, in principle a Tala agent can reason about the need for a new language to describe concepts in some domain, and then create or improve one. Constructions may be used as conceptual structures to declaratively represent the syntax and semantics of a new language, enabling it to be created and improved by conceptual processes. The invention of a new language can then help accelerate learning in a domain, by enabling more effective development of concepts in the domain via induction, abduction, analogy, etc. If a new language is used for cooperation in a domain by Tala agents, then its invention corresponds to creation of a Wittgensteinian 'language game'. (See Vogt's 2005 paper on language evolution in language games. Also see Bachwerk and Vogel (2011) regarding language evolution for coordination of tasks.)

Thus, in principle a Tala agent could demonstrate human-level intelligence in creating new representations. Due to Minsky's identification of the knowledge representation problem (§4.2.6) for baby machines, this is a major strength of the TalaMind approach.

3.7.3   Curiosity

To support higher-level learning, an intelligent system must have another general trait, 'curiosity', which at the level of human intelligence may be described as the ability to ask relevant questions. In maintaining the ontological generality of English, questions expressed in Tala can be about anything, at any level of reasoning or meta-reasoning, in or about any context. A question (not asked rhetorically) has an implication that a Tala agent knows it does not know something. So, to ask a relevant question a Tala agent needs to identify something it does not know, which is relevant to know in or about some context.

Per §3.6.6.1, if a Tala sentence X appears in a context, it represents that X is considered true in that context. Implicitly, if X exists in a context, a Tala agent can also assert "I know X" within that context.

```
(know
        (wusage verb)
        (subj ?self)
        (obj X))
```

A statement of this form might be asserted if it is needed by a conceptual process, and the design of conceptual processing could prevent an infinite loop asserting "I know I know I know ... X". Instead, if it is necessary to represent infinite recursion of self-knowledge of X, a Tala agent could assert:

```
(know
        (<- ?k)
        (wusage verb)
        (subj ?self)
        (obj (and X ?k)]
```

Likewise, if "I do not know X" is asserted in a context then a Tala agent could also assert "I know I do not know X", if needed, in the form:

```
(know
        (<- ?k)
        (wusage verb)
        (subj ?self)
        (obj
                (and
                        (know
                                (wusage verb)
                                (adv not)
                                (subj ?self)
                                (obj X)
                                )
                        ?k]
```

There are several ways a statement of the form "I do not know X" can originate, depending on the context and nature of X, and that corresponding, relevant questions can be created. For example:

- A Tala agent may perceive something in the environment, which it is unable to recognize via processing at its associative and archetype levels, causing the percept at its linguistic level "I do not know what ?x is", where ?x refers to whatever is perceived. This leads to a corresponding question "What is ?x?", the relevance of which may be determined by the environment interaction levels. Relevance itself may be represented by Tala sentences, such as " 'What is ?x ?' is relevant because ?x creates a loud noise."

- A Tala agent may learn about things that are not known in the same way people do, by being told or reading about them. Thus a Tala agent may read "No one knows how to travel faster than light." If the Tala agent does not have any knowledge of its own, then it can accept this as a sentence about its own knowledge, yielding corresponding questions such as "How can one travel faster than light?". The Tala agent may learn about the relevance of such questions from the media that report them, e.g. "The light speed barrier limits our ability to physically explore the universe".

- A Tala agent may also learn that it does not know something, in the process of trying to achieve a goal, and encountering an obstacle it does not know how to overcome. This is also a natural way that people learn they do not know something. The relevance of a question about how to overcome the obstacle then depends on the importance and relevance of the goal.

- Questions may also be created in a more general, undirected fashion, based on observations, goals, etc. Thus, if O is observed, a Tala agent may automatically create questions like "Why was O observed?", "Was O a coincidence?", "What caused O?", "What will O cause?", etc. For any answers to such questions, further questions of the same form may be generated, as young children enjoy proving. The relevance of such questions depends on pragmatic issues: To what extent does an answer have practical bearing on other, relevant questions, to what extent is it possible to answer a question.

In general, both generating questions and reasoning about their relevance can be guided within the TalaMind architecture by mechanisms for reasoning and learning based on abduction, analogy, causal and purposive reasoning, and self-programming. Given the broad generality of these mechanisms, and the ontological generality of Tala, in principle a Tala agent could demonstrate human-level intelligence in identifying and asking relevant questions. Chapters 5 and 6 illustrate how questions can be supported in the demonstration system, showing how a Tala agent could improve its method for making bread by asking a relevant question, "What other features would thick, soft bread have?"

3.7.4   Imagination

The word *imagination* in its common usage refers to the ability to create mental stories, and to mentally construct and simulate ways to solve a problem. Imagination allows us to conceive things we do not know how to accomplish, and to conceive what will happen in hypothetical situations. To imagine effectively, we must know what we do not know, and then consider ways to learn what we do not know or to accomplish what we do not know how to do.

The previous section, on curiosity, discussed how a Tala agent can know what it does not know, relevant to a context. There are several ways a Tala agent can use imagination to satisfy its curiosity, corresponding to the abilities of human-level intelligence:

If a Tala agent does not recognize something perceived in the environment, and humans or other Tala agents don't know what it is, the Tala agent can reason about its purpose and nature, e.g. based on observations of its shape and behavior. This reasoning may involve nested conceptual simulation of how the object might be used or might behave in hypothetical situations, or experimentation in real situations. This reasoning may or may not be successful, but then people are often unable to discern the nature and purpose of a strangely shaped object never seen before.

For other kinds of known unknowns, a Tala agent can reason about ways to answer questions or solve problems, depending on what is unknown and the context. This may involve nested conceptual simulation of hypothetical contexts, or experimentation in real situations. It may leverage the multi-level reasoning and higher-level learning abilities of human-level intelligence, including abductive, analogical, causal, and purposive reasoning. Per Hypothesis III, it can involve processing of mental spaces and conceptual blends — viz. Fauconnier and Turner's (2002) discussion of the Buddhist Monk riddle. This reasoning may or may not be successful, but the point remains that these forms of imagination to resolve unknowns can be supported in principle by the TalaMind architecture.

In addition to imagination for answering questions and finding ways to accomplish what could not previously be done, the word 'imagination' is also used in another sense, corresponding to dreams or daydreams, inventing stories and plays, etc. In principle, a Tala agent could engage in such imagination, using the mechanisms previously discussed for nested conceptual simulation; analogical, causal, and purposive reasoning; conceptual blends, etc. There might be value for a Tala agent to do so occasionally, as a way of generating new ideas. In principle, nested conceptual simulation could simulate multiple levels of stories within stories, or plays within plays, or combinations of these.

3.7.5 Sociality, Emotions, Values

The TalaMind approach supports nested conceptual simulation of hypothetical scenarios by Tala agents, to support reasoning about other agents' perceptions and attitudes in response to one's actions. This provides a starting point toward reasoning about emotions and developing values and social understanding of emotions, exploring research directions that have been investigated by scientists such as Ortony *et al.* (1988), Ridley (1996), Picard (1997), Pinker (2002), Norman (2004), Bach (2009 *et seq.*), Larue *et al.* (2018), and McDuff and Czerwinski (2018).

Lacking human bodies and sensory abilities, perhaps AI systems could never fully appreciate human sympathies. Yet this may enable AI systems to have emotional intelligence (§2.1.2.9). This is a topic for future research in developing the TalaMind approach.

3.7.6 Consciousness

What is theoretically required, to claim a system has consciousness? To claim a system achieves 'artificial consciousness' it should demonstrate:

> *Observation of an external environment.*
> *Observation of oneself in relation to the external environment.*
> *Observation of internal thoughts.*
> *Observation of time: of the present, the past, and potential futures.*
> *Observation of hypothetical or imaginative thoughts.*
> *Reflective observation: observation of having observations.*

This wording is adapted from the 'axioms of being conscious' proposed by Aleksander and Morton (2007) for research on artificial consciousness, discussed in §2.3.4. They used first-person, introspective statements to describe these aspects of consciousness. Rather than referring to subjective awareness, the above statements refer to observations. The term 'awareness' is vague from a computational perspective, while 'observation' can be physically defined.[62]

A Tala agent can perform all the above forms of observation, at least in principle. Observation of an external environment is supported by environment interaction systems, which perform observations that may be expressed as percepts at the linguistic level of concept representation in the TalaMind architecture. Such concepts may represent a Tala agent's observations of itself in relation to the external environment, using sentences with the reserved Tala variable ?self as the object or subject of verbs.

A Tala agent's internal thoughts are concepts expressed as Tala sentences. These are created and processed by conceptual processes. Matching and processing Tala sentences amounts to observing them, within the TalaMind architecture. Such observations may be recorded in other Tala sentences, which may themselves be matched and processed, supporting reflective observations that may be expressed in sentences representing recursive self-knowledge (cf. §§3.7.2.2, 3.6.7.6). The TalaMind architecture allows an agent to have multiple subagents engage in self-talk, communicating in the Tala conceptual language of thought, each referring to the Tala agent by a common reserved variable ?self. This provides another mechanism for reflective observation.

The TalaMind architecture includes contexts representing perceived reality and event memory, supporting observation of Tala sentences in relation to time for the present and past. Conceptual processing may create and match Tala sentences that express hypothetical or imaginative thoughts, which may occur within hypothetical contexts representing potential futures. The Tala language allows representation of concepts that include or refer to concepts, supporting 'higher-order thoughts' within a Tala agent (cf. Rosenthal, 2005).

---

[62] Observation may be considered as a physical process that occurs when one physical system changes state or performs an action, as a result of gaining information about the state of another physical system. Observation is intrinsic to computation, because each step of a computation involves observation of symbols.

In addition to these capabilities of artificial consciousness , there is also Chalmers' Hard Problem, regarding whether a system can have the first-person, subjective experience of consciousness that people have. This will be discussed separately, as a theoretical issue, in §4.2.7.

3.8 Summary

This chapter has analyzed theoretical questions for the hypotheses, and discussed how a system could in principle be designed according to the hypotheses, to achieve higher-level mentalities in human-level AI. It discussed theoretical issues for elements of the TalaMind architecture, and presented affirmative theoretical arguments and explanations for how the TalaMind approach can be developed successfully.

This analysis showed that the TalaMind approach allows addressing theoretical questions that are not easily addressed by other, more conventional approaches. For instance, it supports reasoning in mathematical contexts, but also supports reasoning about people who have self-contradictory beliefs. Tala provides a language for reasoning with underspecification and for reasoning with sentences that have meaning, yet which also have nonsensical interpretations. Tala sentences can declaratively describe recursive mutual knowledge. Tala sentences can express meta-concepts about contexts, such as statements about consistency and rules of conjecture. And the TalaMind approach facilitates representation and conceptual processing for higher-level mentalities, such as learning by analogical, causal, and purposive reasoning; learning by self-programming; and imagination via conceptual blends.

4. Theoretical Issues and Objections

> So it is in contemplation: if a man will begin with
> certainties, he shall end in doubts; but if he will be
> content to begin with doubts, he shall end in
> certainties...For metaphysic...we will take this hold,
> that the invention of forms is of all other parts of
> knowledge, the worthiest to be sought, if it be possible
> to be found. As for the possibility, they are ill
> discoverers that think there is no land, when they can
> see nothing but sea.

~ Francis Bacon, *The Advancement of Learning*, 1605

∞

The previous chapter presented constructive theoretical arguments supporting the TalaMind approach. This chapter addresses theoretical issues and objections for the possibility of achieving human-level AI in principle and for the TalaMind approach.

4.1 Issues and Objections *re* the Possibility of Human-Level AI

4.1.1 Dreyfus Issues

Dreyfus[63] (1992) gave several criticisms of AI research, identified from the 1960s through the 1980s. He identified theoretical issues for human-level AI to address, rather than theoretical objections to its possibility in principle. In discussing the future of AI research, Dreyfus (1992, pp.290-305) left open the possibility that a child machine approach (§1.1) could achieve human-level AI, if his theoretical issues could be addressed – though he was very skeptical about the potential to address these issues, practically.

Russell and Norvig (2010) noted several issues Dreyfus identified have been accepted as design aspects for intelligent systems. These are also important for research in the TalaMind approach, which incorporates responses to Dreyfus' criticisms. For example, this thesis will focus on symbolic processing of conceptual structures, without claiming this is completely sufficient to achieve human-level AI. Other

---

[63] This discussion refers in general to Hubert L. Dreyfus, though some of his work was co-authored with his brother Stuart E. Dreyfus.

technologies may also be needed, such as connectionism or quantum information processing. Likewise, this thesis does not assume the mind operates with a fixed set of formal rules. In the TalaMind approach, executable concepts may be modified by other executable concepts, or accepted as input from the outside environment (analogous to how people can learn new behaviors when given instructions). The approach also includes a variety of other methods, including constructions, semantic domains, mental spaces, and conceptual blends. Nor do Tala executable concepts need to be context-free. Rather, the approach investigates how intelligent systems can understand language and behave intelligently in context-sensitive situations.

In the TalaMind approach, substantial knowledge about the world (including encyclopedic and commonsense knowledge) can be represented in natural language descriptions that can be processed symbolically, but this is not a set of facts each logically independent of all others. Nor is this knowledge claimed to be completely sufficient for human-level AI: rather, an intelligent system will need to acquire some knowledge through interactions with its environment.

Dreyfus (1992, p.56) discussed a "metaphysical assumption" of some AI researchers, that the 'background' of relevant knowledge in a situation is itself an object with its own set of features. He noted that in general, the broadest context or background of any description is open-ended, and potentially infinite. The problems Dreyfus identified with this assumption are essentially those related to representing commonsense knowledge, and related to the qualification problem, i.e. how to avoid specifying an infinity of possible qualifications for a particular action or description – how to describe a process that can be successful when virtually anything could go wrong.

The TalaMind approach is open to use of all the methods AI researchers have developed for the qualification problem (viz. Russell & Norvig, 2010). One possible method is to specify behaviors using generic descriptions and defaults, and to support open-ended, functional reasoning when defaults are not satisfied, without trying to specify everything that might be the case about the environment. For example, in the TalaMind 'discovery of bread' demonstration, a step occurs where one of the Tala agents, Ben, decides to pound some grain, to see if that removes shells from grain. At present, the demonstration does not specify how he pounds the grain, but it could be extended to support Ben reasoning as follows:

```
Ben thinks he needs something harder than a shell to pound
grain, to break a shell.
Ben knows a hammer is a hard tool normally used for pounding.
Ben looks for a hammer, but cannot find one.
Ben looks for something else that is hard and can be used for
pounding.
Ben finds a nearby stick, and uses it to pound grain.
```

Thus a Tala executable concept could specify "find something hard to pound grain" and not have to specify all the details of a context, for the executable concept to be useful for many contexts. The TalaMind approach allows specifying executable concepts very generally, and using or adapting such concepts to support actual, specific contexts, without having to specify all possible conditions of a context. Of course, it's possible a Tala agent does not find a way to apply an executable concept in a particular context.

4.1.2   Penrose Objections

Penrose (1989 *et seq.*) has presented the following claims:

1. Computers cannot demonstrate consciousness, understanding, or human-level intelligence.

2. Some examples of human mathematical insight transcend what could be achieved by computers.

3. Theorems of Turing and Gödel showing theoretical unsolvability of certain logical problems imply human intelligence transcends computers. (Two arguments are given that relate to previous similar arguments presented by Lucas in 1959 and Gödel in 1951.)

4. Human consciousness depends on quantum gravity effects in microtubules within neurons (an hypothesis with Hameroff).

Penrose (1994 – *Shadows of the Mind*) refined and extended arguments presented in 1989 (*The Emperor's New Mind*), so this thesis will focus entirely on his 1994 book and later works.

4.1.2.1   General Claims *re* Intelligence

Penrose's view (1994, pp.37-40) is that one cannot be genuinely intelligent about something unless one understands it, and one cannot genuinely understand something unless one is aware of it. He considers awareness to be the passive aspect of consciousness, with free will being the active aspect. These are commonsense[64] notions of intelligence, understanding, and awareness, and this thesis is broadly in agreement with them. Chapter 2 lists understanding and consciousness as unexplained features of human-level intelligence that need to be supported in human-level AI.

Penrose says intelligence, understanding, and consciousness are real phenomena, worthy and possible to understand scientifically, but he declines to define them, saying his discussion needs to use an intuitive perception of what these terms mean.

This thesis defines human-level intelligence by identifying and describing capabilities not yet achieved by any AI system, which include natural language understanding, higher-level forms of learning and reasoning, imagination, and consciousness. However, this thesis does not claim to give full definitions, nor does it claim to prove computers can achieve human-level intelligence, understanding, or consciousness. Per §1.6, this thesis can at most present a plausibility argument.

Penrose (1994, pp.53-55) explains his general claim that computers cannot achieve human-level intelligence, understanding, and consciousness by saying he does not believe computers can:

- Achieve human-level understanding and visualization of commonsense knowledge about physical objects such as a block of wood, the motions of rope, or the fact that if Abraham Lincoln's left foot is in Washington, it is very likely his right foot is also.

- Achieve human-level understanding of the meanings of words and natural language, for communication with humans, about concepts such as 'happiness', 'fighting', and 'tomorrow', because a computer cannot have human-level awareness of experiences involving such concepts.

This thesis does not focus on visualization of commonsense

---

[64] There are some fine points: Often one becomes conscious of things one does not understand, and one can also be unconscious of how one understands something – these points are important for later discussion of Searle's Chinese Room argument. The topic of free will is outside the scope of this thesis.

knowledge, though it does not appear that there is any in-principle demonstration that it is impossible for computers. (Computer simulation of physical motion in virtual reality is relevant.) Sections 2.2.3 and 3.7.5 discuss understanding of experiences involving embodiment, emotions, etc.

4.1.2.2 Claims *re* Human Logical Insight

Penrose claims some examples of human logical and mathematical insight transcend what could be achieved by computers. The first claim is that a computer could not understand there are an infinity of natural numbers, based on a few examples of how the sequence can be extended by addition. It does not appear there is any in-principle demonstration of this. Rather, it is similar to capabilities Lenat showed in the AM program (Lenat, 1976).[65] It would require the program to include innate concepts for 'equivalence class', 'infinite loop', and 'infinite set'. Given these innate concepts, there is no reason in principle the program could not learn from examples that words like 'zero', 'one', 'two', 'three', 'four' represent different equivalence classes, and that adding 'one' to any equivalence class gives a new, larger equivalence class.

There is also no reason in principle a computer could not have some degree of awareness it was observing and learning these things. It would need to use meta-reasoning to recognize that adding 'one' can give an infinite loop, which it should not repeat forever, and that in principle this implies there are an infinity of numbers. This kind of meta-reasoning and concepts about infinite loops and infinite sets are important for an intelligence kernel to have, to support human-level AI. They are worthwhile and possible in principle.

Regarding human logical insight, Penrose *et al.* (1997, pp.102-106) gives two examples of chess positions, which a little human thought shows that White can play and draw, but for which an expert-level

---

[65] AM started with concepts representing sets, substitutions, operations, equality, etc. It did not have concepts for proof, single-valued functions, or numbers. It simulated discovery of concepts for natural numbers, multiplication, factors, and primes, but it did not have a built-in concept of infinity. Its behavior leveraged built-in features of Lisp (list structures and recursive functions).

chess program playing White may take a rook and lose. These chess problems are examples where human meta-reasoning about chess positions trumps the lower-level search capabilities of conventional chess programs. They don't prove computers in principle cannot employ such meta-reasoning, and approach chess problems the same way a human does. This does not require brute force search, as Penrose suggests. So, these particular examples do not support Penrose's claim that human-level intelligence is in principle non-computable.

Penrose (1994) gives other examples of mathematical proofs that depend on human understanding of geometrical reasoning and mathematical induction. These include commutativity of multiplication, and a proof that a sum of successive hexagonal numbers 1, 6, 12, ... 6n is always the cube of an integer. These are also cases where a computer program could in principle employ the same kind of reasoning, and do not prove human-level intelligence is noncomputable.

4.1.2.3 Gödelian Arguments

The situation becomes more challenging with Penrose's arguments that theorems of Turing and Gödel imply human intelligence transcends computers. For clarity (and following others) these will simply be called Gödelian arguments, since Turing did not agree with them. Similar arguments were presented by Gödel himself, and by Lucas in 1959.

Gödel (1951) argued that either absolutely unsolvable Diophantine problems exist or the human mind is infinitely more powerful than any finite machine. [66] Feferman (2006) analyzed Gödel's argument, and concluded it does not establish anything definitive about the general mathematical capacity of minds and machines.

Lucas (2003) summarized his argument. Since it is along the same lines as the arguments of Gödel and Penrose, the following pages will focus on Penrose's arguments, though first we note that Feferman (2011) discusses Lucas' argument in some detail. To address Gödelian arguments in general, Feferman proposes the representation of human

---

[66] Charlesworth (2016) presents a theorem called the COAT Theorem (Computationalism-impossible Or "Absolute" Truth), which provides an analogous result to Gödel's (1951) conclusion, without assuming some controversial idealizations that Gödel had assumed in the argument he gave in 1951.

mathematical thought as an 'open-ended schematic formal system', for which the language of mathematics is considered to be extensible. This is compatible with the TalaMind approach, which uses an extensible language of thought. Lucas (2011) responds to Feferman, writing that such a system could be open-ended and creative but would not be a deterministic machine. However, the TalaMind approach does not require human-level AI to be deterministic. Rather, one could argue that human-level intelligence must be non-deterministic, in this sense of enabling open-ended, creative development of concepts.

Penrose (1994) presented two arguments claiming the unsolvability of certain logical problems implies that human intelligence transcends computers. His first argument was based on Turing's proof of undecidability for the halting problem. The second was based on Gödel's incompleteness theorems for mathematical theories capable of expressing arithmetic.

Each of the following authors wrote papers critical of Penrose's first and/or second arguments: Chalmers (1995b), Feferman (1995), McCarthy (1995), McCullough (1995), McDermott (1995), and Moravec (1995). Penrose (1996) responded to each of these authors, accepting some of their corrections but disagreeing with their arguments and conclusions. Separately, Davis (1990, 1993) has criticized Penrose's Gödelian arguments – I am unaware of a reply from Penrose to Davis. More recently, Franzén (2005), Lindström (2001, 2006), and Shapiro (2003) have given responses objecting to Penrose's second argument – I am unaware of responses by Penrose.

There the matter appears to stand. Penrose continues to believe he has put forward valid arguments that human intelligence transcends what can be achieved by computers, while several mathematicians and computer scientists disagree. Shapiro (1998) provides an analysis of the debate, which is critical of both sides, and concludes the hypothesis that human-level AI can be achieved is not precise enough to be limited by the incompleteness theorems.

This thesis cannot resolve this debate, involving mathematical logic, issues of philosophy of mind, and representation in human-level AI. Yet having summarized where the argument stands, I will offer some remarks that may be helpful in considering the topic and its potential impact to the quest for human-level AI.

To begin, Turing considered both his and Gödel's theorems as possible objections in his 1950 paper on machine intelligence. Turing noted there is no proof these results do not also apply to human intelligence. This remains as an objection to Penrose's arguments, though Penrose believes human intelligence is exempt because he thinks humans can understand a non-computational, ideal world of mathematical truths.

Penrose (1997, pp.112-113)[67] noted that Gödel allowed it could be possible for a theorem-proving machine to exist that would be equivalent to human mathematical intuition, without it being possible to prove the machine is equivalent, or prove it would always output only correct theorems about finite number theory.[68]

Gödel seems to focus on the unknowability of the limits of human intelligence, as Penrose remarks. In contrast, Turing (1947) focused on the idea that human intelligence may not be limited by mathematical theorems if it is allowed to be fallible. Turing did not claim human intelligence is unknowable or non-algorithmic, but just that it is fallible.

In response to Turing, Penrose (1997) wrote that it was implausible that mathematicians or computers could use unsound procedures to find mathematical truths.

Yet considered generally, human intelligence (including that of mathematicians) is fallible. To require human-level AI to be infallible would be to require it to exceed human-level intelligence. We should not expect that if computers achieve human-level intelligence, then such computers will be infallible, in part because human-level AI will need to process incorrect and conflicting concepts received from humans, or developed by human-level AI. Human-level AI could be subject to many of the same limitations in perception, disambiguation, reasoning, and unintended consequences that affect humans.

It is one of the strengths of human-level intelligence that it can detect and deal with errors and logical inconsistencies, and overcome its fallibility. This is an aspect of robustness, one of the key features of human-level intelligence listed in §2.1.2. In developing human-level AI we should strive to emulate this capability, to enable human-level AI to detect, quantify, and limit errors as much as possible, but it would be a serious mistake to promise infallibility.

---

[67] The reference to Penrose (1997) is to the book *The Large, the Small, and the Human Mind*. Other parts of the book were written by Shimony, Cartwright, Hawking, and Longair. Longair also edited the book.

[68] See also Wang (1996, pp.184-185).

Yet Penrose (1996) asks us to imagine that all the methods of correct mathematical reasoning that humans might use can be contained in a sound formal system. He then compares the performance of Turing machines versus this infallible, unassailable ideal. While he accepts that individual mathematicians can make errors, Penrose says human mathematicians must have some understanding of an ideal 'Platonic world' of mathematical truth that is beyond computation, to aid them in judging whether their ideas are correct and in seeking correct theories. He then asks how can human mathematicians understand this non-computational ideal world of concepts, if humans are just computational systems?

To address Penrose's question, consider his (1994, pp.411-421) discussion of three worlds):

- The *physical world* of reality, as governed by the laws of physics.

- The *mental world*, i.e. the world of human consciousness, thoughts, and perceptions.

- The *Platonic world*, of mathematical truth, ideal mathematical forms, etc.

The mental world emerges from the physical world, perceives the physical world, and constructs theories about laws of physics that govern the physical world. These theories are expressed in mathematics, using 'ideal forms' such as natural and real numbers, geometrical structures, etc. These ideal forms are references to the Platonic world, which seems to emerge from our mental world. Remarkably, we find the physical world conforms to mathematical theories of the laws of physics that predict it with great precision, at every scale. So, the physical world seems to be based on the Platonic world.

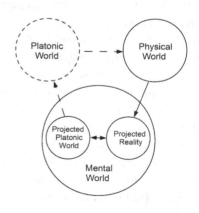

Figure 4-1 Mental Projected Worlds

Penrose's question may be answered as follows: Per Figure 4-1, within the mental world the human mind constructs a '"projected /

perceived reality" of the physical world (viz. §2.2.3 and Jackendoff, 1983). Likewise, we may also construct a projected / perceived reality of a Platonic world, which may be called the "projected Platonic world". This would be a mental representation for our knowledge and understanding of a Platonic world. Initially, our projection of the Platonic world may derive from what we observe in the physical world, via projected reality. For example, our observation of spatial relationships in the physical world enables us to create a projected Platonic world that includes theories of geometry. Using concepts from our projected Platonic world, we may create theories to predict our projection of the physical world. Remarkably, our projection of the physical world obeys some of our theories very well, and we find we can measure and predict it with great accuracy using these theories. Yet to the extent our concepts in the projected Platonic world were derived from our observations of the physical world, it may be less remarkable that the physical world is consistent with itself, and conforms to theories based on projected Platonic concepts.

In any case, we may think and act as though the Platonic world really exists, independent of our projection of it, just as we think and act as though the physical world exists independent of our projected reality, which it does. Indeed, we tend to think and act as though we directly perceive the physical world, and may likewise tend to think we directly perceive a Platonic world, even though we are only thinking and acting relative to projections we construct in our mental world.

All of this may help guide mathematicians and physicists in their quests for theories, help them to judge theories, correct errors, etc. We may construct proofs, judge correctness and 'mathematical beauty' or symmetry, etc. relative to our perception of a projected Platonic world. However, it does not guarantee our projected Platonic world is correct, just as we are not guaranteed our projected reality is correct. We are subject to misconceptions about both. Occasionally, we may need to radically adjust our understanding of physical reality, and accordingly we may need to radically adjust our understanding of the Platonic world. When this happens, mathematical ideas previously considered 'un-Platonic' may need to be accepted into the projected Platonic world. We might need to consider that parallel lines may eventually meet, or eventually diverge.

As for whether the Platonic world really exists, separate from our projected Platonic world, and separate somehow from the physical world, pragmatically it may be impossible to test an answer to this question: An answer might not make any conceivable difference to us in our abilities to predict the physical world.[69] On the other hand, there are philosophical arguments for the existence of the Platonic world, e.g. the Quine-Putnam indispensability argument (viz. Colyvan, 2011). It's beyond the scope of this thesis to consider this question further.

Whether the Platonic world really exists or not, there is no reason in principle why a human-level AI could not have an internal conceptual construct, providing a projected Platonic world, making it a computational entity with access to projections of ideal concepts. These projections could be finite, discrete computational representations, even if the ideal concepts are non-computable: For example, the concept of a countable infinite set can be represented by a finite sentence stating that a non-terminating computation will generate all elements of the set. The concept that a set is uncountable can be represented by a finite, counterfactual sentence stating that if there were a countable list of all elements in the set, then a diagonalization would construct a new element of the set, not in the list.

Penrose (1994, pp.147-149) also considers whether natural selection could have selected for mathematical ability, and suggests it is more likely to have selected for a general ability to understand – this does seem to be what has actually happened: Humans do have a general ability to understand, though few demonstrate advanced mathematical ability. To illustrate this, Penrose (1997, p.113) shows a drawing of a mathematician pondering geometrical diagrams while a sabertooth tiger is about to pounce on him, as other people in the distance focus on more practical matters for survival, like farming, constructing dwellings, or hunting mammoths, noting these tasks involve knowledge that is not limited to mathematics. In considering whether human mathematical intelligence can be described by an algorithm, Penrose asks how such an

---

[69] The quote at the beginning of this chapter is apropos. Bacon continued: "But it is manifest that Plato, in his opinion of ideas, as one that had a wit of elevation situate as upon a cliff, did descry that forms were the true object of knowledge; but lost the real fruit of his opinion, by considering of forms as absolutely abstracted from matter, and not confined and determined by matter; and so turning his opinion upon theology, wherewith all his natural philosophy is infected."

algorithm could have been created by natural selection.

The answer may be given by translating this into the question "How did natural selection give us, or how do we develop, a projected Platonic world?"

An answer to this question is that nature selected for understanding spatial relationships and spatial forms, for us to successfully perform actions in the spatial, physical world. Natural selection also selected for linguistic concept representation and logical reasoning, for us to communicate and understand causal relationships and cooperate in hunting mammoths, building houses, and growing crops. Also, nature would presumably have selected for an ability to detect and avoid infinite loops in our own reasoning, to recognize and transcend logical paradoxes, so we did not become transfixed by inconsistencies in our thinking when sabertooths were nearby.

Thus natural selection could have given us the conceptual precursors that allow us to develop projected Platonic worlds in our minds. This explains why, as Penrose notes, a child is easily able to grasp the concept of an infinity of a natural numbers, after learning the concept that 'adding one' always creates a new natural number. And it explains how more complex mathematical concepts can be incorporated into a student's projected Platonic world, through teaching or independent discovery.

In sum, what is at issue in Penrose's Gödelian arguments is the extent to which computers can recognize and transcend logical paradoxes, to reason about theories as well as within them, in the same way people do. In his response to Penrose, McCarthy (1995) made this point.

Human mathematicians can follow the arguments of Turing and Gödel, and reason about formal systems to recognize certain problems are unsolvable within them, without becoming trapped in infinite loops trying to solve the problems. Turing and Gödel's unsolvability arguments are essentially examples of meta-reasoning, i.e. reasoning about what happens within logical systems to detect logical contradictions and potential infinite loops. There is no reason in principle why computers cannot also use meta-reasoning of this sort. This is an aspect of metacognition, one of the key features that Chapter 2 lists for human-level intelligence.

4.1.2.4   Continuous Computation

Buescu, Graça, and Zhong (2011) study the extent to which mathematical problems in dynamical systems theory can be solved by symbolic computation, and show that solutions to some problems are in general not computable by Turing machines. It is tempting to suggest that issues related to noncomputability might be avoided with continuous (and potentially chaotic) computation, using infinite precision real numbers. On the one hand, research on "hypercomputation" and "super-Turing machines" provides arguments that various forms of continuous computation can transcend Turing machines (Copeland, 2002).

On the other hand, there are issues with the physical realizability of such machines. Davis (2006) argues hypercomputation must be considered a myth. Penrose (1994, pp.21-26, 177-179) considers and discounts continuous computation and chaos theory as possible explanations for non-computability of human intelligence and consciousness. And if continuous computation can be physically realized, then in principle it should be usable by human-level AI, i.e. it does not appear to be theoretically restricted to natural intelligence. This is an interesting topic, but one outside the scope of this thesis.

4.1.2.5   Hypothesis *re* Orchestrated Objective Reduction

Penrose and Hameroff (2011) proposed a theory called Orchestrated Objective Reduction, or Orch OR, which conjectures that human consciousness depends on non-computable quantum gravitational effects in microtubules of neurons. Microtubules are nanoscale cylindrical lattices of a protein called tubulin, which they conjecture can act as biomolecular quantum computers.

It is out of scope for this thesis to discuss Orch OR in detail, since it involves topics in theoretical physics and neurobiology. Here it will simply be noted that Orch OR is a controversial hypothesis. This thesis will therefore take a neutral stance on Orch OR, and only consider the implications if it is correct.

If microtubules actually are performing computations that support human-level intelligence achieved by the brain (which at this point is unproved), then the information processing capacity of the human brain could be orders of magnitude larger than would be estimated otherwise. Penrose and Hameroff (2011, pp.9-10) estimate potentially 10^8 tubulins in each neuron switching at 10^7 times per second, yielding potentially

10^{15} operations per second per neuron. This would give each neuron a processing capacity one tenth what could otherwise be estimated for the entire brain (10^{11} neurons, 10^3 synapses per neuron, 10^2 transmissions per synapse per second = 10^{16} operations per second). And this would not count the potential combinatorial effect of quantum parallelism in microtubule computations.

Such calculations should be considered very preliminary, astronomical upper bounds on what might be achieved. Yet they indicate quantum processing at the level of microtubules might add enormous computational power to the brain, if it is happening.

However, this would not necessarily make human-level AI unobtainable. Penrose and Hameroff (2011) suggest that much of the microtubule processing in the brain may be allocated to perception (vision, hearing, touch, taste, etc.) rather than to cognition and consciousness. [70] Thus, microtubule computation might challenge AI most in areas that depend on massive perceptual processing, and perhaps not challenge AI as much in higher-level mentalities. There could still be a prospect that higher-level mentalities can be achieved using symbolic, discrete processing of conceptual structures, e.g. via the TalaMind approach.

Further, the Orch OR theory does not imply artificial systems could not also use massive nanoscale quantum computation. Penrose (1994, pp.393-394) appears to agree with this, and it would be within the spirit of Turing's 1950 paper to allow nanoscale quantum computation in achieving human-level AI. Turing wanted to allow any engineering technique to be used in a machine. He limited his conjecture to conventional digital computers in part to exclude human cloning as a technique that by definition would artificially create human intelligence. Yet Turing, like Penrose, was interested in quantum processes (viz. Hodges, 2011).

---

[70] They write that only tens of thousands of neurons are considered to be involved in a human brain's cognition and consciousness at a time. In calculating a 40 Hz frequency (25 msec period) of Orch OR events to correspond with EEG gamma synchrony, they consider 1% of the microtubules in 20,000 neurons might be involved in consciousness, for a total of 2×10^{10} microtubules.

4.2   Issues and Objections for Thesis Approach

4.2.1   Theoretical Objections to a Language of Thought

As noted in §2.2.1, the language of thought hypothesis is controversial and has perhaps not yet been widely accepted, because of philosophical arguments pro and con, e.g. concerning issues such as whether an innate language is needed to learn an external language and the degree to which an innate language must contain all possible concepts, or constrains the concepts that can be learned and expressed.

However, as noted previously, there is an elegant argument that concepts must be expressed as sentences in a mental language: Since natural language sentences can describe an effectively unlimited number of concepts, and the brain is finite, concepts must be represented internally within the mind as structures within a combinatorial system, or language. (Viz. Jackendoff, 1992, pp.23-24.)

This provides a theoretical license for the TalaMind approach to include a conceptual language as a language of thought. Jackendoff's argument does not preclude the use of a syntax based on a particular natural language to support representing and processing conceptual structures for the semantics of natural language.

4.2.2   Objections to Representing Semantics via NL Syntax

Sections 3.3 through 3.6 present affirmative theoretical arguments that in principle natural language syntax can be used to represent natural language semantics, and provide theoretical explanations for how to do so. This section further addresses specific theoretical objections.

4.2.2.1   The Circularity Objection

Some critics may object that using a mentalese with a syntax based very largely on the grammar of English to help represent the semantics of English seems circular (cf. §2.2.1). However, an impression that it seems circular is not a theoretical reason in principle why natural language syntax cannot be used by a conceptual language to support an AI system. Nor is such an impression correct. The TalaMind approach does not involve an infinite regression in representing syntax or semantics (cf. §4.2.8). Meaning may involve referent pointers (§3.6.1) to non-linguistic levels of Figure 1-1.

4.2.2.2   Objection Syntax Is Insufficient for Semantics

Section 3.3 gave an argument that it is theoretically possible to represent natural language semantics using natural language syntax, but it expressly did not claim that all semantics can be represented by syntax. Nor does the TalaMind approach rely entirely on syntax to represent semantics. In addition to referent pointers (§3.6.1), the semantics of Tala conceptual structures also derives from how their syntax is processed by TalaMind conceptual processes in a conceptual framework, and how such processing supports a Tala agent's interactions with its environment and other agents in the environment. The TalaMind approach does not claim that syntax is sufficient to represent semantics, only that some syntax is useful (even arguably necessary) for representing semantics. That is why natural languages have syntax, after all, and it is why an internal, conceptual language needs a syntax. (See also §§4.2.4, 4.2.8.)

4.2.2.3   Ambiguity Objections to Natural Language

It may be objected that natural language sentences are ambiguous, and the inferences a natural language sentence allows depend on which of its interpretations are considered. Therefore, it may be argued that one of the requirements for a conceptual language should be that it is able to unambiguously represent specific interpretations of natural language sentences.

As discussed in §3.6.3.7, people often reason with only partly disambiguated natural language sentences, using an interpretation that is specific to an extent appropriate in a given context. A Tala agent must do the same or else it will not converse with people in a way they consider achieves human-level intelligence. Tala supports representing and reasoning with natural language expressions that are ambiguous, while providing the flexibility to remove ambiguity when needed.

Though Tala is based on a natural language, it augments natural language to support removing ambiguities in several ways, discussed in §3.6.3: Tala's representation of natural language syntax can remove syntactic ambiguity. Lexical ambiguities can be removed with Tala (wsense) and (wreferent) expressions. Other forms of ambiguity, such as individual/collective, specific/non-specific, can be removed by (wsense) semantic annotations for determiners, e.g. expressing a usage of "all" as either (wsense individual) or (wsense collective). Ambiguities related to coreference can be removed using pointers and the (<- ?p) notation for binding pointers.

A specific form of the ambiguity objection is the following: If a Tala representation of an English sentence has multiple possible interpretations then (according to §3.6.3) in general more specific interpretations may be represented by other Tala sentences. Yet if an ambiguity is resolved by replacing the sentence by another, typically more complex sentence, then that sentence will have its own ambiguities, with the danger of an infinite regress.

This argument may be addressed with three remarks: Again first, we are not forced to remove all ambiguities, and indeed need to support reasoning with partly disambiguated sentences. Second, §§3.6.3 and 3.2.1 allow exceptions to the method of representing interpretations via other sentences in Tala. We are not forced into an infinite regress if some other formal language or notation can represent an interpretation more concisely or accurately than is possible in Tala, or if the ambiguity can be resolved by semantic annotation of the Tala sentence.

Third, the meanings of some linguistic concepts depend on references to lower levels of concept representation in Figure 1-1. An embodied, situated, intelligent agent may ascribe meaning based on observations of its environment. Other symbols and concepts may be given meaning procedurally within an agent by virtue of how they affect and are processed by conceptual processing.

For many concepts we can at best choose to believe we agree about their meanings. For higher-level concepts there may always be some ambiguity and uncertainty. Human-level AI will have the same challenge in this respect as human intelligence.

4.2.2.4   Objection Thought Is Perceptual, Not Linguistic

It may be objected that the TalaMind approach cannot achieve human-level AI because human thought is not linguistic in nature, rather it is perceptual. This is an extension of the claim that human-level intelligence requires human physical embodiment, discussed in §2.2.3. For example, Lakoff and Johnson (1999) wrote that a purely syntactical approach to language could not be successful, isolated from semantics, emotion, context, etc. Similarly, Evans and Green (2006) referred to the research of Barsalou (1993 *et seq.*) who described *perceptual symbols* as perceptual states stored in long-term memory that function as symbols.

However, Barsalou (2012) discussed how perceptual symbols may be used in the human conceptual system and concluded that successful future theories of functional conceptual systems are likely to integrate

GOFAI[71] approaches, connectionism and other statistical approaches, and systems for simulating modality-specific knowledge about the body and environment.

The approach of this thesis is in the direction Barsalou recommends. The three conceptual levels shown in Figure 1-1 correspond to the three accounts of the human conceptual system he describes. Although this thesis takes a direction at the linguistic level different from GOFAI for reasons explained in §1.2, it is open to using traditional AI methods as needed, per §3.2.1. The TalaMind approach is also open to perceptual symbols in the lower levels of Figure 1-1, integrated with the linguistic level.

Barsalou (1999) said computers could implement perceptual symbol systems yet expressed doubt that computers can achieve human-level intelligence, due to a lack of sensory-motor systems needed to represent human concepts. This appears to be a concern based more on the state of robotics technology than an in-principle theoretical objection. In either case, §2.2.3 addresses such concerns through its discussion of virtual embodiment.

To summarize, this thesis holds that thought is both linguistic and perceptual in nature, and that neither can be totally sacrificed in achieving human-level artificial intelligence. While internally both humans and computers use physical symbols and processes without being aware of their meaning or even existence, at higher levels of representation both humans and Tala agents can in principle ascribe meaning to words and actions, and have semantic originality (§3.7.2.2).

4.2.3   Weizenbaum's Eliza Program

The TalaMind approach, representing natural language semantics based largely on the syntax of a particular natural language, may encounter theoretical objections invoking Weizenbaum's (1966) Eliza computer program (viz. Fauconnier & Turner, 2002). Eliza could interact with humans via text messages, and simulate a psychotherapist. People could type simple English sentences as input to Eliza, and Eliza would

---

[71]GOFAI = Good Old-Fashioned Artificial Intelligence (Haugeland, 1985). Incidentally, Barsalou's research is sometimes described as investigating a 'Perceptual Symbol Systems Hypothesis.' This hypothesis does not appear to have been formally stated as such, but is in effect described by Barsalou (1999).

generate English sentences in response, using simple parsing and substitution of key words into standard phrases. Users who didn't know how Eliza worked sometimes thought the program was intelligent. Here is a dialog I created with a re-implementation of Eliza[72]:

```
>> Hello.
How do you do. Please state your problem.
>> I need to finish my thesis on understanding natural
language.
Your thesis on understanding natural language  ?
>> Yes, I need to explain how you work.
You seem to be quite positive.
>> Well, I'm certain that you don't understand me.
Is it because you are certain that I don't understand you
that you came to me ?
>> Yes, very much so.
You are sure.
>> Well I do know how you work.
Perhaps in your fantasies we do know what each other.
>> I do fantasize that someday my own program may understand
me.
Why do you say your own program may understand you  ?
>> My program will be based on cognitive linguistics and AI
theory.
Does that suggest anything else which belongs to you ?
>> I need to get back to writing my thesis now.
Is it important that your thesis now  ?
>> Goodbye.
Goodbye. Thank you for talking to me
```

The theoretical objection critics may raise against the TalaMind approach would be to say it is just a glorified version of Eliza, and that like Eliza a Tala agent will just manipulate English sentences without understanding them. The response to this objection is that the TalaMind approach is very different from Eliza, and does have the potential for Tala agents to understand English sentences, unlike Eliza. In the TalaMind approach, a Tala agent will use encyclopedic and commonsense knowledge to reason about the meanings of English sentences. A Tala agent will have a conceptual framework that includes a perceived / projected reality, giving it 'virtual embodiment'. The TalaMind approach will support learning and discovery by reasoning analogically, causal and purposive reasoning, meta-reasoning, imagination via conceptual simulation, and internal dialog between

---

[72] The re-implementation of Eliza was written in Java by Charles Hayden, using the script of patterns and responses provided in Weizenbaum's (1966) paper. Hayden's program is available at http://www.chayden.net/eliza/Eliza.html, as of January 31, 2019.

subagents in a society of mind (§2.3.3.2) using a language of thought. It will support semantic disambiguation, natural language constructions, metaphors, semantic domains, etc. Thus the TalaMind approach has the potential to emulate understanding of English sentences to a degree that humans may eventually say indicates human-level intelligence, even when humans understand how the TalaMind system works.

4.2.4   Searle's Chinese Room Argument

The TalaMind approach, representing natural language semantics based largely on the syntax of a particular natural language, must confront theoretical objections based on Searle's "Chinese Room" philosophical thought experiment. It has been the subject of unresolved debate since 1980, though the philosophical issues are complex enough that people on both sides may believe they resolved it in their favor, long ago. Cole (2009) provides a survey of this debate.

Searle and other proponents of the Chinese Room argument have claimed it shows human-level intelligence requires human physical embodiment, or at least the embodiment of mental processes within the human brain. Section 2.2.3 gives reasons for believing that human physical embodiment in general is not needed to understand natural language.

The perspective of this thesis has been developed independently, though elements of it have been previously proposed by others. People on both sides of the Chinese Room argument have noted it depends on the nature of consciousness and understanding. This thesis will discuss these topics, but will not trace the relationships of its perspective to the large number of previous responses by other thinkers.[73,74]

---

[73] This thesis will not rely on what Searle says is the most frequent reply he receives and does not accept, which he calls the "Systems Reply" – Russell & Norvig (2010) support the Systems Reply, noting that others including McCarthy and Wilensky have proposed it. I agree with their arguments, and with the arguments given by Chalmers (1996), but present a different argument to discuss how consciousness and understanding interact within the mind.

[74] See also Mc Kevitt & Guo (1996) for a discussion addressing the Chinese Room by representing meanings spatially and pictorially. From a cognitive linguistics perspective their approach appears equivalent to

Searle asks us to imagine a person placed in a room who understands English but does not understand Chinese. The person has instructions written in English that tell him how to process sentences written in Chinese. Through a slot in the room he receives pieces of paper with Chinese sentences on them. He follows his instructions in English to process the Chinese sentences and to write sentences in Chinese that he pushes through the slot out of the room. Searle asks us to consider the person in the Chinese Room as equivalent to a computer running a software program, and to agree that neither the Chinese Room and its contents nor the person inside it using English instructions understands Chinese, but that the person inside the room understands English. From this, Searle argues that no computer running a program can truly understand a natural language like English or Chinese.

To an outside observer, it appears the Chinese Room (or the person inside it) understands Chinese, and it also appears the Chinese Room understands English, since if English sentences are put on paper pushed into the room, meaningful English replies are received from the room. So for the outside observer, the Chinese Room satisfies the Turing Test for Chinese and English. However, the person inside the room does not rely on the Turing Test. He 'knows' that he does not understand Chinese and 'knows' that he does understand English. It is important to ask:

Precisely how does he know these things?

Answering this question involves a discussion of how consciousness, knowledge, and understanding interact within a mind. As discussed in §§2.3.4 and 3.7.6, consciousness includes the ability of a person to observe his thoughts, i.e. to observe in his own mind what he thinks or knows. To the extent that a person can observe his or her thoughts, much of the understanding process appears to happen seamlessly, and the person may only be conscious of the higher-level results of the process. Referring to the discussion of understanding given in §2.2.2, a person can be conscious that he or she is able or unable to understand meanings, yet not consciously perceive all the interpretants constructed in the mind, nor all the conceptual processing of interpretants that

---

image schemas (Johnson, 1987) or perceptual symbols (Barsalou, 1993 *et seq.*).

constitutes the understanding process.

From this perspective, the person in the Chinese Room knows he does not understand Chinese because he is conscious that he is not able to create interpretants for the Chinese characters, i.e. he is conscious (observes in his own mind) that he does not know what the Chinese symbols represent.

He knows that he does understand English because whenever he reads an English sentence he can observe at least indirectly in his own mind that interpretants are created representing its meaning. He is conscious (observes in his own mind) that he has an understanding of what the English sentence refers to. Thus the person in the room is conscious of the fact that he understands English, and conscious of the fact that he does not understand Chinese. Yet he is not fully conscious of how he understands English, i.e. what process he uses to understand English.

Searle's argument does not preclude the possibility that the person may subconsciously process symbols to understand English in essentially the same way that he consciously processes symbols to emulate understanding Chinese. The person may use a program expressed in his innate mentalese to support processing English subconsciously, in the same way that he uses the external instructions written in English to support processing Chinese consciously. He may have constructed his internal program for understanding English when he learned how to understand English as a young child, and now be executing the program subconsciously. Thus we normally learn how to do new things consciously, and later perform complex processes unconsciously after they become routine. So from this perspective, Searle's argument does not prove that symbol processing cannot constitute understanding of semantics.

In Searle's 1980 paper, he discounted this possibility, on the grounds that it would appear "incredible" to claim or suppose that a program could have the same input-output performance as a native human speaker or to assume that human speakers can be described as program instantiations. In effect, Searle allowed his doubts regarding the potential for computer programs to describe human-level intelligence and enable computers to understand natural language, to support his conclusion that they cannot do so. In his 1980 paper, Searle apparently did not fully consider that the person in the Chinese Room might unconsciously understand English in the same way he consciously emulates understanding Chinese. Thus, Searle's Chinese Room argument does not disprove the potential value of the TalaMind approach, nor of the quest for human-level artificial intelligence in general.

This discussion of how consciousness interacts with natural language understanding is relevant to understanding in general. Much of what we perceive and do happens automatically and unconsciously, with consciousness being drawn to things we do not understand, perceptions that are anomalous, actions and events that do not happen as expected, etc. Once we become conscious of something anomalous, we may focus on trying to understand it, or trying to perceive it correctly, or trying a different action for the same purpose. (Cf. Whitehead's 1929, p.161, statement that consciousness is involved in the perception of contrast between an erroneous theory and a fact.)

Searle (1992) gave a second theoretical argument against Strong AI, contending that computation can only be ascribed to physical processes when they are observed by human intelligence. Searle appears not to consider that physical observation and causality are intrinsic and essential to computation. While the arrangement of symbols on a tape is syntactic, Turing's definition of computation also describes physical observation (reading a tape) and physical actions (moving relative to the tape, and writing on it), as well as state changes within a machine. This carries over to modern computers, though they use other physical representations for symbols, and other physical processes than Turing described. Physical observation, actions, and causality occur within a computer, even if the system is not conscious or intelligent. So, a computer can perform computations independently of whether it is observed externally by human observers. Further, a computer could in principle observe its computations and ascribe meaning to them, if it is computationally reflective (§3.7.2.2). Chalmers (1996) gives a more extensive refutation of Searle's arguments.

4.2.5   McCarthy's Objections to Natural Language Mentalese

McCarthy (2008) wrote a paper on child machines, called *The well-designed child*. In general what he wrote appears compatible with this thesis, except for his arguments (p.2009) that natural languages like English would not work as languages of thought, both for humans and for AI systems with human-level intelligence. He said it would be "appropriate" for a robot's language of thought to be based on logic.

From the perspective of this thesis, McCarthy was mistaken in discounting natural language as a basis for an AI system's language of thought. His reasons indicate an incorrect assumption that the characteristics of external, public spoken or written natural language would necessarily obtain for an internal, private language of thought with a syntax based on natural language. Since McCarthy (2008) directly contradicts Hypothesis II of this thesis, the following paragraphs respond to his arguments.

McCarthy noted that the brain processes information in parallel, and said our thoughts don't occur to us as linear English expressions. The Tala conceptual language has a syntax that corresponds to natural language, but this syntax is not expressed within a purely linear string of words. Instead Tala sentences are expressed as tree-like list structures. If necessary, such structures could be processed in parallel by multiple conceptual processes, and multiple conceptual processes could reason in parallel about the same structure, or different parts of a structure.

McCarthy said the brain must use something equivalent to pointers to refer to sensations, but that English has to use word expressions, since we can't use English to give each other pointers into our brains. He noted that a robot's logical language could use pointers to locations in its memory or to its senses.

However, Tala allows pointers to be used within conceptual structures based on natural language. Pointers are valid since they only need to be dereferenced within a Tala agent using the mentalese to refer to its own concepts, and mental or physical states. The Tala mentalese could include reserved variables to serve as pointers to states of a Tala agent's senses or body. This would, incidentally, address questions posed by Wittgenstein (1953) regarding how an agent could use a 'private language' to refer to personal sensations such as pain, which other agents cannot directly perceive. This does not mean that a Tala agent would need to use a single numeric value to represent pain or other sensations, or qualia. As McCarthy suggested by referring to the visual cortex, an agent could describe or refer to the state of a complex field of sensory data.

McCarthy's claim that our thoughts do not occur to us as lengthy English expressions is based on introspection, which may not be valid: How we perceive we think may not really be how we think. In any case, if we think of a thought as a 'chunk' rather than a long sentence, we may do so by using a pointer to the thought as a whole, independently of the structure of the thought. The thought might still have a complex structure, with a syntax and semantics that correspond to natural language.

Likewise, his argument that a language of thought must function much faster than speech does not preclude use of a mentalese with syntax and semantics that correspond to natural language. We may process mentalese conceptual structures internally much faster than we can translate such structures into speech and perform physical speech acts. Natural language supports representing thoughts about processes that are ongoing, and Tala pointers could point to such processes. (The TalaMind demonstration system includes 'process objects' representing active executable concepts.)

McCarthy noted that human brains evolved from animal brains, which don't use natural languages, and said some of our thoughts must be close to the thoughts of evolutionary ancestors. This argument discounts the possibility other animals may have very simplified languages of thought, which apparently remains open to future research by neurobiologists. In any case, this thesis is not focused on how animals or humans think, but on how machines might emulate human thought, and what languages machines might use for emulating thought.

Emulation of mental capabilities may be accomplished without exact duplication or replication of the supporting physical processes, languages, etc. Thus, we know that human brains have different physical processes than electronic computers, and no human brain operates using the machine language of any electronic computer. So, if one is going to argue that predicate calculus and logic are adequate as an internal language for a robot achieving human-level AI, then one should be willing to consider the possibility that a symbolic formalism[75] based on natural language may also be adequate.

Deciding which kind of language is "appropriate" for a robot to use ultimately depends on what works best for representing the thoughts needed within a system having human-level artificial intelligence. Formal logic languages do not easily represent the broad range of thoughts we can express with natural language (§2.3.1).

---

[75] For clarity, the term 'logical formalism' has been replaced here by 'symbolic formalism'.

McCarthy's (2008) claim that meaning has greater importance than grammar discounts a major purpose of syntax, which is to help represent meaning. So, this claim is not a valid argument against using natural language as a basis for design of a language of thought.

McCarthy observed that a language of thought might be reorganized as a child develops, saying this might explain why most people cannot remember much from infancy. This does not preclude that a child may have an innate language of thought and perhaps internalize English as a language of thought, reorganizing or extending the innate language.[76] Nor does it preclude that a child machine could use a natural language like English as its innate language of thought, and perhaps learn and internalize other natural languages.

McCarthy (2008) wrote that his paper was prompted by Pinker's (1994) book *The Language Instinct,* which supports the idea that humans have a language of thought but is not specific about its nature (cf. §2.2.1). In saying English and other natural languages would not be suitable for an AI system's language of thought, McCarthy (2008) did not cite any previous research investigating this idea for AI systems.

4.2.6   Minsky's Issues for Representation and Learning

Minsky (2006, p.178-182) was not optimistic about prospects for the 'baby machine' approach to human-level AI. He said previous research efforts toward general-purpose learning systems failed because systems stopped being able to extend themselves. He attributed this to the inability of systems to develop new representations for knowledge.

The nature of representation is discussed throughout this thesis. The thoughts of Peirce and Wittgenstein were considered in §2.2.2. Representations for natural language semantics were discussed in §3.6, including Fauconnier and Turner's theory of conceptual blends (§3.6.7.9). Smith's (1982) issues for representation will be discussed in §4.2.8.

Per Hypothesis III, the TalaMind approach uses methods from cognitive linguistics to support multiple levels of representation. Two major levels are the linguistic and archetype levels of the TalaMind

---

[76] See §2.2.4 regarding natural language and inner speech. Indeed, the fact that individuals do not remember much that happened before they first learned spoken languages tends to support natural language playing a role in thoughts and memories.

architecture. Within the linguistic level, multiple levels of abstraction, domains of thought, and reflection can be represented using Tala, a natural language of thought for AI. Learning and representing knowledge about new domains using analogies and metaphors with previously known domains is an aspect of higher-level learning, illustrated in the TalaMind demonstration system (§6.2.1, §6.3.3).

In principle, a Tala agent will be able to represent and discover concepts that can be expressed via natural language, and even invent new languages to represent concepts (§3.7.2.3). Also, the TalaMind architecture is open to inclusion of iconic mental models for spatial-temporal representation and reasoning (§1.5, §2.3.6), open to formal logic and mathematical representations (§2.3.1), and open to associative representation of concepts that can be learned and perceived by neural networks (§1.5). Thus, TalaMind has been designed to have the abilities of human-level intelligence for creating new representations. It should address Minsky's representation issue for baby machines – if it does not, then we may learn something new about representations.

Minsky (2006) identified three other problems for baby machines related to optimization, complexity, and investment.

Regarding optimization and complexity, he observed that if a system is optimized to perform well, then changes to the system may have a greater chance of degrading its performance. It can become more challenging to improve the system, or for the system to self-improve. Likewise, as a system becomes more complex, there will be a greater chance of changes having unforeseen consequences.

If we consider any complex system and imagine making a random change to it, then the odds are the change will harm or disable the behavior of the system. However, well-designed systems are modular and hierarchical. They have limited, well-defined interfaces between different components. Each module has a specific, well-defined function within the system. Natural language or formal sentences may describe global and local characteristics of the modules, their interfaces, and interactions. Someone who understands these descriptions of the system design can be successful in improving the system by making changes that improve how a module performs its function, without changing its interface to other modules, at any level of the hierarchy. This enables human intelligence to overcome Minsky's complexity and optimization issues – though we must still allow time for debugging unexpected problems. There is no reason in principle why the TalaMind approach cannot also support this ability of human-level intelligence (cf. §2.3.3.6.2 and Doyle, 1980, pp.33-38).

Regarding investment, Minsky observed that as a system becomes more optimized there is a tendency to invest less in alternatives. This is simply an economic fact of life, though it is very important: If a system works well in performing its function, then the time and costs necessary to identify and implement an improvement may prevent doing so. Properly viewed, higher-level learning by human intelligence is an economic process[77]: An intelligent system must meta-reason about any new learning endeavor, to decide whether it is economically worthwhile to spend time thinking about how to improve or invent a system to achieve some purpose. Higher-level learning may be considered as an *"economy of mind"* (Wright, 2000), an extension of a generalized society of mind (§2.3.3.2).

In principle the TalaMind approach can support higher-level learning processes related to economic considerations. The use of a natural language mentalese will enable a TalaMind system to represent and meta-reason with concepts about the difficulty and value of making an improvement to a method, as well to represent and reason about ways to improve a method, etc.

4.2.7 Chalmers' Hard Problem of Consciousness

The "Hard Problem" of consciousness (Chalmers, 1995a) is the problem of explaining the first-person, subjective experience of consciousness. For this thesis, there is the theoretical issue of whether a Tala agent can have this first-person, subjective experience. This is a difficult, perhaps metaphysically unsolvable problem because science relies on third-person explanations, based on observations. Since there is no philosophical or scientific consensus about the Hard Problem, this thesis may not give an answer that will satisfy everyone. On the other hand, the TalaMind approach is open to different answers for the Hard Problem:

- If the answer to the Hard Problem is that the subjective experience of consciousness is a byproduct of neurobiological processes in the human brain, as suggested by Searle, then this

---

[77] Thus, the economic aspects of improvements and innovation were central to Adam Smith's work, and even more so to the work of Joseph Schumpeter, and modern-day economists.

thesis would agree with Chalmers that a human-like conscious experience could be implemented by a computation (Chalmers, 1996, p.315), and argues that the TalaMind approach can theoretically provide the right computational approach. There does not appear to be sufficient reason to accept Searle's claim that only neurobiological processes can produce consciousness.

- If the answer to the Hard Problem is that the subjective experience of consciousness is a byproduct of quantum information processing in the brain, as suggested by Penrose and Hameroff (2011), then the TalaMind approach is open to inclusion of quantum information processing, if necessary. However, §3.7.6 did not invoke quantum processing to describe how third-person aspects of consciousness could be supported in the TalaMind approach, and this author is not yet convinced that quantum processing is needed for the subjective, first-person experience of consciousness.

- If the answer to the Hard Problem is that the subjective experience of consciousness is a byproduct of non-symbolic information processing in the brain, e.g. connectionism, then the TalaMind approach is open to inclusion of non-symbolic processing, if necessary. While this thesis discusses support of third-person consciousness via symbolic processing, perhaps other aspects of consciousness, such as fringe consciousness, may benefit from connectionism, holography, or other technologies – this would be a topic for future research. However, this author is not yet convinced that non-symbolic information processing is needed for the subjective, first-person experience of consciousness.

- If the answer to the Hard Problem is that the subjective experience of consciousness is an illusion, as suggested by Dennett and Blackmore, then the TalaMind approach could include conceptual processes that would simulate and report having such an illusion, if it is useful or important to do so. Blackmore (2011, p.285) notes this answer could also be given for other approaches to artificial consciousness.

Observation is intrinsic to symbolic computation, and the theoretical requirements for TalaMind related to consciousness are stated in §3.7.6

in terms of observation. Another answer to the Hard Problem, consistent with the TalaMind approach, is that there is a first-person aspect inherent to observation: An observer encounters an observation from a first-person perspective, and others can only discuss the observer's observation from a second- or third-person perspective. This does not imply every physical system performing an observation is conscious, because most physical systems do not observe themselves, observe internal concepts, etc. A simple physical system like a thermostat is not conscious by these criteria, even though it performs physical observations of temperature. Yet a Tala agent performing all the observations required for consciousness discussed in §3.7.6 would encounter each of these observations from a first-person perspective. Arguably, this would give the Tala agent an intrinsic first-person experience of its consciousness, which others could only observe indirectly. Though with current technology that experience could not have the sensory richness of human consciousness, it could still be argued from a theoretical perspective that a first-person experience of consciousness exists.

As Dennett and Blackmore each note, our perceptions of consciousness are to some extent illusions, things that exist or happen but are not the way they seem. Thus we dynamically piece together our visual perception of the environment, perceiving a continuous unity in space and time out of saccadic eye motions. The perception that we have a single, unitary self may also be an illusion, considering evidence from split-brain experiments. Blackmore (2011) discusses Libet's neurophysical experiments indicating that unconscious actions can precede conscious perceptions of intentions to perform the actions. If consciousness is an illusion, it may be an illusion that perceives itself, and an illusion that it can perceive itself.

A Tala agent can have multiple subagents engage in self-talk, communicating in the Tala conceptual language of thought, each referring to the Tala agent by a common reserved variable ?self (viz. §§3.6.7.13, 5.4.16). At least from a logical standpoint, this provides a representation of a single experiencer, composed of subagents.

The TalaMind approach, including support for a projected reality within the conceptual framework, does not imply a homunculus within a 'Cartesian Theatre', leading to a problem of infinitely nested homunculi. Rather, within a Tala agent's society of mind (§2.3.3.2) two subagents can engage in mental dialog, though more are permitted (the prototype demonstration system has three subagents within each Tala agent). The construction and use of a projected reality do not require infinite recursion, beyond what can be finitely represented, e.g. via circular pointers.

No discussion of consciousness would be complete without some mention of philosophical arguments regarding "zombies". A philosophical zombie is a hypothetical system with behavior indistinguishable from humans, yet which does not have consciousness, understanding or intelligence (cf. Searle, 2004, p.93). This concept is introduced to support a philosophical argument that consciousness does not logically follow from observations of external behavior. The issue is not relevant to the TalaMind approach, which requires us to observe the internal design and operation of the system, and to consider to what extent its internal conceptual processing supports aspects of consciousness. Because we are not confined to external behavior, TalaMind systems are not philosophical zombies.

4.2.8 Smith's Issues for Representation and Reflection

Smith (1982) noted the lack of an accepted, theoretical definition of the term 'representation', writing:

"there is remarkably little agreement on whether a representation must 're-present' in any constrained way (like an image or copy), or whether the word is synonymous with such

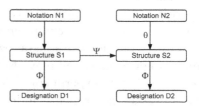

Figure 4-2 Semantic Mapping Functions

general terms as 'sign' or 'symbol'...further confusion is shown by an inconsistency in usage as to what representation is a relation between...Thus a KLONE structure might be said to represent Don Quixote tilting at a windmill; it would not be taken as representing the fact or proposition of this activity." [78]

Since 1982, published research on conceptual graphs, description logic, ontology representation languages, etc., appears to have taken

---

[78] Quotations and Figure 4-2 in this section are used with permission of Brian C. Smith and MIT Press.

essentially a propositional approach. However, Smith's definitional issues remain worthy of discussion in this thesis on representation and computation of meaning. The term 'representation' has multiple senses, each important in the TalaMind approach.

One sense is synonymous with 'sign' or 'symbol'. The discussions in §2.2.2 and Chapter 3 can be summarized by adapting a diagram from Smith (1982, p.62) in Figure 4-2. It shows relationships between three kinds of semantic functions. θ is a function mapping external sentences (notations) into internal conceptual structures. Φ is a function mapping conceptual structures into their designations in the world. Ψ is a function mapping conceptual structures into other conceptual structures corresponding to their interpretations or implications, within a Tala agent's conceptual framework. Smith writes:

> "As an example to illustrate [Figure 4-2] suppose we accept the hypothesis that people represent English sentences in an internal mental language we will call mentalese … If you say to me the phrase "a composer who died in 1750" and I respond with the name "J. S. Bach", then, in terms of the figure, the first phrase, *qua* sentence of English, would be N1; the mentalese representation of it would be S1, and the person who lived in the 17th and 18th century would be the referent D1. Similarly, my reply would be N2, and the mentalese fragment that I presumably accessed in order to formulate that reply would be S2. Finally, D2 would again be the long-dead composer; thus D1 and D2, in this case, would be the same fellow."

Though his wording in this excerpt is suggestive, Smith does not appear to have discussed the idea that mentalese could itself be based on the syntax of a natural language, such as English. For the TalaMind architecture, conceptual structures S1 and S2 are of course expressions in the Tala mentalese. External notations N1 and N2 are typically written or spoken English sentences, but may be any kind of external percept that a Tala agent represents as a conceptual structure. However, because a Tala agent can only indirectly refer to the external world, the designations D1 and D2 are also conceptual structures in the agent's conceptual framework. Thus, a Tala agent's conceptual structures representing J. S. Bach would correspond to an entry for the long-dead composer in its encyclopedic knowledge. If in a different example, D1 represented a physical person present in the external environment, then it would be a percept in the Tala agent's perceived reality, i.e. again a conceptual structure in the agent's conceptual framework. Finally, if N1 were the phrase "Newton's third law of motion", then its designation D1 would be a Tala conceptual structure corresponding to "For every action there is an equal and opposite reaction", again stored in the agent's encyclopedic knowledge.

This approach addresses the inconsistency in usage that Smith noted concerning "what representation is a relation between", i.e. whether a conceptual structure represents an activity in external reality, or a statement of fact or proposition about an activity. Consider Smith's example:

```
Don Quixote is tilting at a windmill.
(tilt
     (wusage verb) (tense present) (aspect continuous)
     (subj
          ("Don Quixote"
               (wusage noun)
               (naming proper)))
     (obj
          (windmill
               (wusage noun)
               (det a)]
```

Whether this Tala sentence represents (designates) an activity in external reality or is just a proposition depends on the epistemic mode (§3.6.7.1) of the context in which it occurs within the conceptual framework of a Tala agent, which affects how the sentence is conceptually processed by the agent. If the sentence occurs as a percept in the Tala agent's perceived reality context, then it represents the agent's perception of an activity in its environment, and may be conceptually processed as a statement of fact.[79] If it occurs in the Tala agent's event memory for perceived reality, then it represents a memory of an earlier percept. If it occurs in a hypothetical mental space or scenario context for nested conceptual simulation, then it represents a hypothetical activity in that context, effectively a hypothetical proposition. If the sentence occurs in a Tala agent's encyclopedic

---

[79] However, its processing as a statement of fact about perceived reality may be modulated by other knowledge about what is happening in reality. For example, if the Tala agent happens to be watching a play about Don Quixote, then the agent may understand that the sentence represents what is currently happening in the play and that an actor is portraying Don Quixote.

knowledge for a scene in Cervantes' novel about the "ingenious gentleman" of La Mancha, then it is a proposition describing a fictional event, i.e. a statement of fiction.

Because Tala sentences incorporate the syntax of English, such distinctions can be expressed directly within them. Thus, we may have a Tala sentence corresponding to "Cervantes wrote that a fictional character named Don Quixote tilted at windmills." This is a statement of fact, a true proposition about reality, which a Tala agent may have as encyclopedic knowledge. Although it is a statement of fact, it specifically refers to a fictional character and event.

Section 3.7.2.3 discusses how the TalaMind approach and architecture can support a second set of meanings for 'representation', related but not synonymous to semantic mappings between signs or symbols. This is the notion that a representation may be a particular way of describing a problem or situation, or multiple problems and situations. This could range from a particular set of sentences, to a notation or language. It is important for a Tala agent to be able to flexibly develop such representations, to address Minsky's knowledge representation problem for achieving human-level AI (§4.2.6).

Finally, Smith (1982) discusses a third meaning of the term 'representation':

> "If nothing else, the word 'representation' comes from 're' plus 'present', and the ability to *re-present* a world to itself is undeniably a crucial, if not *the* crucial, ingredient in reflective thought. If I reflect on my childhood, I re-present to myself my school and the rooms of my house; if I reflect on what I will do tomorrow, I bring into the view of my mind's eye the self I imagine that tomorrow I will be. If we take 'representation' to describe an *activity*, rather than a *structure*, reflection surely involves representation..."

This meaning of representation is also open to support within the TalaMind architecture, if we take the "mind's eye" of a Tala agent to mean whatever conceptual processes are currently active, and such processes have the ability to recall or imagine (simulate) spatial images. Also, it could be logically equivalent to recall or imagine (conceptually simulate) and process a collection of Tala sentences rather than spatial images. The notion of a mind's eye in reflective thought overlaps previous discussions of observation within artificial consciousness (viz.

§§3.7.6 and 4.2.7).

Some additional theoretical issues are found in six general properties that Smith (1982, pp.42-81) reasoned should be exhibited by any reflective system. Following is a summary of these properties, and discussion of how they are supported in the TalaMind approach and architecture:

1) Reflection "matters", i.e. it is causally connected with behavior. The result of reflecting can affect future non-reflective behavior. Prior non-reflective behavior is accessible to reflective contemplation. A system can create continuation structures that can trigger reflection at later moments.

This property is an aspect of the active side of consciousness, which some authors have equated with freedom of will, though it is more well-defined and easier to discuss. All of these causal connections between reflection and behavior can in principle be supported within the TalaMind architecture: Executable concepts can access and conceptually process (reflect upon) event memory of prior behavior. As a result, such reflective executable concepts can create new executable concepts that affect future non-reflective behavior, or which trigger reflection at later moments.

2) Reflection involves self-knowledge, as well as self-reference, and knowledge is theory-relative.

This property is also supported in the TalaMind architecture. Each agent has a reserved variable ?self for use in concepts representing knowledge about itself, and in executable concepts for reflection. Such concepts can exist within theories.

3) Reflection involves an incremental "stepping back" for a system to process structures that describe its state 'just a moment earlier', and avoids a vicious circle of thinking about thinking about thinking...

This property is also supported within the TalaMind architecture, via the ability for executable concepts to access event memory. There is no need for reflection to involve an infinite, vicious circle of thinking about thinking ... However, it is a feature of the Tala language that it supports finite representations of infinitely recursive concepts (§3.6.7.5).

4) Reflection allows a system to have fine-grained control over its

behavior. What was previously an inexorable stepping from one state to the next is opened up so that each step can be analyzed and changed in future.

This property is also supported in the TalaMind architecture. Since executable concepts can be analyzed by other executable concepts, each step of an executable concept can be analyzed and changed in future behavior.

5) Reflection is only partially detached from what is reflected upon, and is animated by the same fundamental agencies and processes.

This is also the nature of reflection within the TalaMind architecture. Much as Smith's 3-Lisp reflective procedures were still written in 3-Lisp and animated by its processing, Tala reflective executable concepts would also be expressed in Tala, animated by TalaMind conceptual processing.

6) The ability to reflect must be built into a system and its language. Rather than simply having a model of itself, a system must be able to directly analyze and change itself.

This is also possible in the TalaMind architecture, consistent with the TalaMind hypotheses. Executable concepts can directly analyze and change executable concepts, as discussed in §§3.7.2.2, 6.3.3.2.

4.3 Summary

This chapter discussed theoretical issues and objections for the TalaMind approach, or against the possibility of achieving human-level AI in principle. No insurmountable objections were identified, and arguments refuting several objections were presented. These pages have discussed the theoretical issues for AI identified by Dreyfus and philosophical arguments against AI, including Searle's Chinese Room argument and the Gödelian arguments of Penrose and Lucas. I also discussed McCarthy's objections to natural language as a mentalese, Minsky's issues for representation and learning, Chalmers' 'Hard Problem' for consciousness, and Smith's issues for representation and reflection. Considering all these discussions, it does not appear to me that anyone has shown human-level AI is impossible in principle, nor that anyone has shown the thesis approach cannot succeed in principle.

5. Design of a Demonstration System

There is no mode of action, no form of emotion, that we do not share with the lower animals. It is only by language that we rise above them, or above each other—by language, which is the parent, and not the child, of thought.

~ Oscar Wilde, *The Critic as Artist*, 1891

∞

5.1 Overview

Chapter 3 analyzed how a system could in principle be designed according to the TalaMind hypotheses, to achieve the higher-level mentalities of human-level intelligence. It discussed theoretical issues for elements of the TalaMind architecture. This chapter presents a design for a prototype demonstration system, in accordance with the analysis of Chapter 3.

The purpose of the prototype is to illustrate how the thesis approach could support aspects of human-level AI if the approach were fully developed, though that would need to be a long-term effort by multiple researchers. Per §1.6, this thesis cannot claim to actually achieve human-level AI. Hence the demonstration system cannot be claimed to actually achieve the higher-level mentalities of human-level intelligence, it can only illustrate how they may eventually be achieved. This illustration will involve functioning code in a prototype system, but it can only be a small step toward the goal. This distinction is further discussed in Chapter 6.

Likewise, the purpose of the prototype design is not to show the best way to design a system having a TalaMind architecture, better than other possible designs. That is a topic for future research, since it will involve issues of efficiency and scalability, as well as issues of detailed design regarding completeness and generality. Such issues will be very important and challenging. However, if one can make workable design choices for a prototype, this may suggest possibilities for a more full and scalable implementation of TalaMind. Some of the design choices for the prototype may carry over to future systems, though many will not.

5.2   Nature of the Demonstration System

The demonstration system is a functional prototype in which two Tala agents, named Ben and Leo, interact in a simulated environment. Each Tala agent has its own TalaMind conceptual framework and conceptual processes. Each Tala agent uses the Tala conceptual language as its internal mentalese for communication between subagents in its society of mind (§2.3.3.2). Ben and Leo can communicate with each other, and can also perform actions and perceive objects and events in the simulated environment.

The simulation uses the Tala language to represent actions, percepts, and communication between Tala agents. The simulation displays actions, events, and communication between agents using English sentences, which are generated from Tala mentalese expressions, but the agents do not themselves parse English sentences as linear strings of symbols. The demonstration focuses entirely on conceptual processing using the Tala mentalese. The simulation can also display the internal thoughts (Tala conceptual structures) of each Tala agent as English sentences.

Thus, to the human observer, a simulation is displayed as a sequence of English sentences, in effect a story, describing interactions between Ben and Leo, their actions and percepts in the environment, and their thoughts.[80] The story that is simulated depends on the initial concepts that Ben and Leo have, their initial percepts of the simulated environment, and how their executable concepts process their perceptions to generate goals and actions, leading to further perceptions and actions at subsequent steps of the story. The story is 'scripted' in the sense that this author has written executable concepts that work

---

[80] This form of scripted story simulation is different from other research approaches to automatic story generation, e.g. Meehan (1981), Turner (1994), Perez y Perez & Sharples (2004). Also, the system does not use scripts as described by Schank to sequence the actions of an agent throughout a typical situation such as dining at a restaurant. In the TalaMind prototype, each simulation step involves a Tala agent processing different executable concepts. Executable concepts may guide the actions of an agent across time intervals, e.g. to support 'discovery loops', and so in principle could support Schank's scripts.

together to produce the story, to illustrate how Tala agents could perform different kinds of higher-level concept processing.

For the simulations created to date, the stories have involved situations in which Ben is a cook and Leo is a farmer. Two stories have been developed, one about Ben and Leo discovering how to make bread from wheat, and another about Ben and Leo exchanging wheat for bread. Consequently, some examples discussed in this chapter may have references to wheat or bread, or to events in the stories. The stories are discussed in detail in Chapter 6.

The TalaMind prototype demonstration system is written in JScheme, a Scheme dialect of Lisp implemented in Java – some lower-level code is written in Java. The prototype does not integrate external corpora, nor does it support reasoning with large amounts of encyclopedic or commonsense knowledge.

5.3 Design of Conceptual Language

This section presents a design for the syntax of the Tala conceptual language. This syntax is fairly general and flexible, and covers many of the issues discussed by Hudson (2010) for Word Grammar dependency syntax. Such coverage is described to suggest that a Tala syntax could be comprehensive for English, since §1.6 identified this as an issue for the adequacy of the Tala syntax. However, developing a comprehensive Tala syntax for English is itself a very large effort that could occupy multiple researchers.[81] The following pages identify topics for future work and there are probably several other ways the Tala syntax can be improved.[82]

Moreover, it should be noted that creating a comprehensive Tala syntax for English is not a prerequisite for success of the TalaMind approach. It is only necessary that Tala include sufficient syntax to enable representing the general, extensible semantics of English, and to support an intelligence kernel's implementation of higher-level mentalities. Other variations in English syntax could later be added into

---

[81] Thus, Hudson notes that comprehensive treatments of English grammar have spanned 1,000 to 2,000 pages each (citing Quirk *et al.*, 1985; Biber *et al.*, 1999; Huddleston & Pullum, 2002; Carter & McCarthy, 2006).

[82] Pinker's (2014) discussion illustrates some of the virtues of Huddleston & Pullum's (2002) Cambridge Grammar for English.

Tala via constructions, [83] if the syntax and semantics for Tala constructions are sufficiently general (cf. §§3.6.3.13, 5.5.4). Accordingly, the prototype simulations described in Chapter 6 need only use a subset of this section's syntax to illustrate how the TalaMind approach can support the higher-level mentalities of human-level intelligence.

Further, Vogt's (2005) research on emergence of compositional structure suggests that if an intelligence kernel starts with less rather than more Tala syntax for English, then this could be advantageous for learning both syntax and semantics via interaction with English speakers in a real-world environment. This may be a promising direction for future research, though it would also be a very large effort that could occupy multiple researchers.

5.3.1   Tala Syntax Notation

The syntax for Tala is presented using a modified Backus-Naur Form (BNF) notation:

```
:=          means "is defined as"
|           means "or"
< >         are used to surround category names
?           means that the item to the left can appear
            zero or one times
*           means that the item to the left can
            appear zero or many times
+           means that the item to the left appears one
            or more times
{}          grouping symbols for syntax expressions
""          used to quote a symbol or sequence of symbols
;           prefixes a comment, not part of the syntax
```

In this notation, (and) are terminal symbols of the Tala language, while { and } are used as grouping symbols of the modified BNF syntax metalanguage. Parentheses are terminal symbols because (per §3.5.2) a sentence in the Tala language will be a Lisp list structure that represents the syntactic structure of a corresponding English sentence. For example:

```
({<adj-word> | <noun-word>}+)
```

refers to a sequence of one or more adjectives or nouns that is enclosed in parentheses, such as:

```
(large gray steel oil pump piston)
```

---

[83] Constructions could be added either manually by human researchers, or using machine learning techniques.

Of course, this example expression is not a sentence in Tala, nor does it correspond to a sentence in English. It only shows how parentheses as terminal symbols describe list structures. Also in examples throughout this thesis, a right square bracket] stands for as many right parentheses as are needed to close all open left parentheses, using a convention of Lisp programming environments.

5.3.2 Nouns

The Tala syntax for a noun is:

```
<noun> := <common-noun>|<pronoun>|<gerund>|<infinitive>|
          <c-conj-noun>
<common-noun> :=
   (<noun-word> ;a word defined in the Tala lexicon as a noun
       (wusage noun)
       (det <det>)?
       (number {singular | plural | mass})?
       (agreement-number {singular | plural})?
       (agreement-person {I | non-I})?
       (naming proper)?
       <prep-link>*
       <adjective>*
       (comp {<noun>|<c-conj-noun>})?
       (subj-of <verb>+)?
       (obj-of <verb>+)?
       (indirect-obj-of <verb>+)?
       <pointer-bind>?
       (wsense <pointer>+)? ;viz. §3.6.1
       (wreferent <pointer>+)? ;viz. §3.6.1
       )
```

By convention in this thesis, if a Tala syntax rule specifies a list then after the first element specified by the rule, any order may be used for other elements. So, after a word is specified at the start of a noun expression, any of the other elements (wusage, det, number, …) may occur in any order.

Since the Tala syntax is essentially a dependency grammar, a common noun may contain links to other nodes, corresponding to a noun expression. Each link corresponds to one of the slots in the noun expression, as follows:

- det – specifies an optional determiner for a noun.
- number – if not specified, by default the noun is singular. The syntax can express that a noun is plural or that it is used as a mass noun and not counted (e.g. "furniture", "oxygen"), in which case its agreement-number is singular.
- agreement-number and agreement-person – These are used to

support English subject-verb agreement, according to the approach described by Hudson (1999). This is discussed in §5.3.12.2.

- `naming` – The syntax can express that a noun is being used as the name of a particular individual (`proper`). Otherwise by default it is a general noun usage.
- `prep-link` – one or more prepositions can be specified as dependent on a noun.
- `adjective` – one or more adjectives can be specified as dependent on a noun.
- `comp` – optionally allows specifying compound noun expressions.
- `subj-of` – optionally allows specifying the noun is the subject of a verb
- `obj-of` – optionally allows specifying the noun is the object of a verb
- `indirect-obj-of` – optionally allows specifying the noun is the indirect object of a verb.
- `pointer-bind` – optionally allows binding a Tala pointer to a noun.

Compound nouns are not supported by treating nouns as adjectives, since adjectives can be modified by adverbs and it would be ungrammatical to say "the extremely player piano". Hence Tala provides a `comp` link to construct compound nouns. A conjunction of nouns may be used syntactically in place of a noun.

Examples:

```
the whole grain very thin flat bread
(bread
    (wusage noun)
    (det the)
    (adj
        (flat
            (wusage adjective)
            (adj
                (thin (wusage adjective)
                    (adv very)
                    ))))
    (comp
        (grain (wusage noun)
            (adj whole]
```

```
the salt and pepper shakers
(shaker (wusage noun)
     (number plural)
     (comp
          (and (wusage c-conj)
               (salt (wusage noun))
               (pepper (wusage noun))
               ))
     (det the]

the picture Leo painted
(picture
     (wusage noun)
     (det the)
     (obj-of
          (paint (wusage verb)
               (tense past) (aspect perfect)
               (subj
                    (Leo (wusage noun)
                         (naming proper]

the artist painting the picture
(artist (wusage noun)
     (det the)
     (subj-of
          (paint (wusage verb)
               (tense present) (aspect continuous)
               (obj
                    (picture (wusage noun)
                         (det the)]
```

Other information about a noun would be stored in the Tala lexicon or encyclopedic knowledge if necessary, e.g. whether the noun typically indicates someone of a particular gender. It does not appear such information needs to be stored within each Tala sentence using an English noun, though it could be added into the syntax, if necessary, in future research.

An infinitive may also be used syntactically as a noun, with syntax specified in the next section. The syntax for a gerund supports using present participles of verbs as nouns:

```
<gerund> :=
     (<verb-word> ;a word defined in the lexicon as a verb
          (usage noun) ;here being used as a noun,
                         ;in its participle form
          (det <det>)?
          (tense present)
          (aspect continuous)
          (number {singular | plural})
          (agreement-number {singular | plural})
          (agreement-person non-I)
          <prep-link>*
          <adjective>*
          (comp {<noun>|<c-conj-noun>})?
```

```
<pointer-bind>?
    )
```

This syntax allows specifying determiners, adjectives, and creating compound expressions using gerunds, and plurality for a gerund as if it were a noun. Thus it could support a noun expression like "the trainer's frequent runnings and occasional winnings of the Kentucky Derby". Note: A sentence like "Eating this cake is easy" can be represented as the verb expression "Eating this cake" being the subject of the verb expression "is easy", using syntax in the next section.

5.3.3   Verbs

The Tala syntax for a verb is:

```
<verb> := <infinitive> |
    (<verb-word> ; a word defined in the Tala lexicon
                ; as a verb
        (wusage verb)
        (tense {present | past | future})
        (subj-number {singular | plural})?
        (subj-person {I | non-I})?
        (passive)?
        (aspect {simple | perfect | continuous})?
        (modal <modal>)?
        <prep-link>*
        <adverb>*
        (subj <verb-subj>)?
        (obj <verb-obj>)?
        (indirect-obj <verb-obj>)?
        (subj-of <verb>+)?
        (obj-of <verb>+)?
        (sentence-class
            {statement | question | exclamation |
            imperative })?
        (silent)?
        (speechform <word>)?
        (speechorder ovs)?
        (wsense <pointer>+)? ;viz. §3.6.1
        (wreferent <pointer>+)? ;viz. §3.6.1
        <pointer-bind>?
        )

<verb-subj> := <verb-obj> :=
        <noun>|<verb>|<adjective>|<conjunction>
<modal> := can | may | would | should | could...
```

Since the Tala syntax is essentially a dependency grammar, a verb will contain links to other nodes corresponding to a verb expression. Each link corresponds to one of the slots in the verb expression, as follows:

- tense – indicates a present, past, or future tense of the verb.

- `subj-number` and `subj-person` – support English subject-verb agreement, according to the approach described by Hudson (1999). This is discussed in §5.3.12.2 below.
- `passive` – if specified, indicates the verb is performed on the subject, rather than performed by the subject. If not specified, the verb is performed by the subject, i.e. the voice is active by default.
- `aspect` - If not specified, then `simple` by default.
- `modal` – indicates uncertainty or conditionality of the verb.
- `noun-mod` – supports compound verbs in which a noun describes a verb.
- `subj` – indicates subject of the verb.
- `obj` – indicates object of the verb.
- `indirect-obj` – indicates indirect object of the verb.
- `subj-of` – indicates the verb is the subject of another verb or verbs.
- `obj-of` – indicates the verb is the object of another verb or verbs.
- `sentence-class` – indicates whether the verb is a statement, question, exclamation, or imperative.

The syntax allows adverb and preposition dependencies within a verb expression, as well as subject and object dependencies. A verb can be the object or subject of another verb. By choosing combinations of `tense` and `aspect` values, the Tala syntax can specify:

- `present simple` – "see"
- `past simple` – "saw"
- `future simple` – "will see"
- `present perfect` – "have seen"
- `past perfect` – "had seen"
- `future perfect` – "will have seen"
- `present continuous` – "am seeing"
- `past continuous` – "was seeing"
- `future continuous` – "will be seeing"

In English, auxiliary (helping) verbs are used to express most of these combinations, but the Tala syntax can represent them without auxiliary verbs. Use of a modal supports expressions such as "may be seeing". To represent the following combinations in Tala requires nesting of verb expressions as objects of auxiliary verbs:

- present perfect continuous – "have been seeing"
- past perfect continuous – "had been seeing"
- future perfect continuous – "will have been seeing"

The syntax above can support several forms of compound verbs. For example, "started reading" can be represented with "reading" being the object of "started". This could also support a 'stretched verb' like "get rid of X from Y", with "rid" being a verb nested in "get" – conceptual processing is responsible for treating the stretched verb as a combination of two verbs.

Similarly, phrasal verbs combining verbs plus prepositions are supported. The preposition can be treated as dependent on the verb, without a prepositional object, and the verb can be treated as having a direct object. Conceptual processing is responsible for treating a phrasal verb as a combination of the verb and preposition (cf. §3.6.8's discussion of prepositions), leveraging idiomatic information in the Tala lexicon or encyclopedic knowledge.

A slot could be added into the verb syntax to support representing compound verbs in which nouns modify verbs, e.g. "We water and sand blasted the sculpture." However, typically these might be expressed as hyphenated verbs, or as compound words like "sandblast". Since this borders on morphology, it is left for future research.

Some slots in a verb expression are used at present only to control the display of the expression in demonstration output (FlatEnglish):

- `silent` – Allows specifying that the verb is not displayed, to support hidden/silent verbs in expressions, e.g. "He made Madison [to be] secretary of state."
- `speechorder` – By default is `subject-verb-object`, but allows specifying `object-verb-subject`.
- `speechform` – Allows displaying a verb using a different word. For example the verb might be an internal primitive `has-part` and the `speechform` might be `has`.

The `sentence-class` slot affects whether a sentence is displayed ending with ".", "?", or "!", and also affects conceptual processing of a sentence, in the prototype demonstration.

The Tala syntax for infinitive verbs is:

```
<infinitive> :=
    (to (wusage prep)
        <bare-infinitive>
        )

<bare-infinitive> :=
    (<verb-word>
        (usage verb)
        (tense present)
        (aspect simple)
        (subj-number singular)
        (subj-person non-I})
        <prep-link>*
        <adverb>*
        (wsense <pointer>)? ;viz. §3.6.1
        (wreferent <pointer>)? ;viz. §3.6.1
        <pointer-bind>?
        )
```

This syntax for infinitives allows split infinitives (per Huddleston & Pullum, 2005, p.206). The "to" in an infinitive is sometimes described as a particle rather than a preposition, but Huddleston and Pullum (2005, p.144) say that particles are prepositions, with a few exceptions.

5.3.4   Prepositions

The Tala syntax for a preposition is:

```
<prep-link> :=
    (<prep-word> ;a word defined in the Tala lexicon
              ; as a preposition
        (wusage prep)
        <prep-object>?
        <pointer-bind>?
        <c-conj-prep>* ;viz. coordinating conjunctions,
                     ; §5.3.9.1
        <prep-link>*
        (wsense <pointer>)? ;viz. §§3.6.1, 3.6.8
        (wreferent <pointer>)? ;viz. §§3.6.1, 3.6.8
        )

<prep-word> := to | for | from | of | after | when | ...
<prep-object> := <noun>|<verb>|<adjective>|<prep-link>
```

Examples:

```
government of, by and for the people
(government
    (wusage noun)
    (of (wusage prep)
        (people
            (wusage noun)
            (det the)
            (<- ?p)
            )
```

```
(and (wusage conj)
     (by (wusage prep) ?p)
     (for (wusage prep) ?p)
     ]
```

```
She went to Phoenix on Thursday and Tucson on Friday.
(go (wusage verb)(tense past)(aspect perfect)
    (subj she)
    (to (wusage prep)
        (Phoenix
              (wusage noun)
              (naming proper)
              (on (wusage prep)
                  (Thursday (wusage noun) (naming proper))
                  ))
        (and (wusage conj)
             (to (wusage prep)
                 (Tucson
                       (wusage noun)
                       (naming proper)
                       (on (wusage prep)
                           (Friday (wusage noun)
                                   (naming proper]
```

```
dressed in red
(dress
     (wusage verb)
     (tense past)
     (in (wusage prep)
         (red
               (wusage adj)]
```

```
The curtain fell after the fat lady sang.
(fall
     (wusage verb)
     (tense past) (aspect perfect)
     (subj
           (curtain (wusage noun) (det the))
           )
     (after (wusage prep)
            (sang
                  (wusage verb)
                  (subj
                        (lady (wusage noun)
                              (det the)
                              (adj fat)
                              ]
```

The object of a preposition is optional, at least for some prepositions, e.g. one can say "I put the bread in the oven and left it in."

The approach taken in this thesis is to allow a word to be used as either a preposition or a subordinating conjunction, much as a word may be used as either a noun or verb. Thus we could use "if" as a preposition in "one if by land", but use "if" in a subordinating / structured "if-then-else" conjunction in an executable concept (§5.3.9.2).

Of course, the TalaMind approach is not dependent on this; Tala could accommodate other linguistic analyses.

5.3.5   Pronouns

The standard English pronouns may be used in place of nouns:

```
<pronoun> := <pron>|<pron-poss>|<pron-det>|
                <pronoun-quest>|<pronoun-exp>
<pron> : = ({I | you | he | she | it | we | they …}
<pron-poss> : = ({mine | yours | his | hers | ours | theirs}
<pron-det> : = { this | these | those | that | any | some |
                all }
<pronoun-quest> : = { who | why | how | when | where |
                    what | which}
<pronoun-exp> :=
        ({<pron>|<pron-poss>|<pron-det>|<pronoun-quest>
        (wusage pronoun)
        (number {singular | plural})?
        (agreement-number {singular | plural})?
        (agreement-person {I | non-I})?
        {subjective | objective}?
        (wsense <pointer>)? ;viz. §3.6.1
        (wreferent <pointer>)? ;viz. §3.6.1
        <pointer-bind>?
        )
```

Agreement number and person are used to support inflections according to the approach described by Hudson (1999), discussed in §5.3.12.2. Some determiners may be used by themselves as pronouns, e.g. "this", "that", etc. The syntax allows these to also be specified with agreement number and person to support inflections. However, this syntax takes a different approach from Hudson (2010), who argues that all determiners should be considered as pronouns. The above syntax also supports words that may be used as subordinating conjunctions or as interrogative pronouns (e.g. "how", "why", "when", "where", etc.).

5.3.6   Determiners

The Tala syntax for determiners is:

```
<det> := <pure-det>|<n-det>|<pron-det>|<det-quest>|
        <poss-pron-det>| <clitic-poss-det>|<nested-det>

<pure-det> : = {the | a | an | some | no}
<n-det> := <number>
<pron-det> : = { this | these | those | that | any |
                some | all | each …}
<poss-pron-det> : = { my | your | his | her | its | their}
<det-quest> : = { what | which}
<clitic-poss-det> :=
        ("'s"
            {<noun> |
            (<c-conj>
```

```
                  <noun>
                  <noun>+
                  )
          })
  <nested-det> :=
      (<det>
          (det
              {<det>|<nested-det>}
              (of <wusage prep>)?
              ))
```

Some determiners may be used as pronouns as discussed in the previous section. Possessive pronouns are also determiners. Some determiners are interrogative (e.g. "what", "which"). So we can represent expressions like:

> the box, or the boxes
> some box, or some boxes
> this box, or these boxes
> his box, or his boxes
> which box, or which boxes

The potential for future support of number agreement in determiners and complements is discussed in §5.3.12.1. The syntax above allows numeric counts as determiners, though the syntax for specifying numbers is open for definition. At a minimum it could include integer numbers (e.g. "35"), or it could include English number-word strings, e.g. "thirty-five". This is also left as a topic for future work. Thus at a minimum, one could have a Tala expression like:

```
five boxes
(box
      (wusage noun)
      (number plural)
      (det 5)
      )
```

The Tala syntax above supports the clitic "'s" for creating possessive determiners from nouns and compound nouns (cf. Hudson, 2010):

```
John's house
(house
      (wusage noun)
      (det
          ('s (John (wusage noun) (naming proper)))
          ]

John and Mary's house
(house
      (wusage noun)
      (det
```

```
           ('s
               (and (wusage c-conj)
                    (John (wusage noun) (naming proper))
                    (Mary (wusage noun) (naming proper))
                    ]

the room and board's daily cost
(cost (wusage noun)
     (det
          ('s
               (and (wusage c-conj)
                    (room (wusage noun) (det the))
                    (board (wusage noun))
                    )))
     (adj daily)
     )
```

There does not appear to be a need for Tala syntax to support the clitic "'s" as an abbreviation for the verb "is". It could be added into the syntax if needed.

Finally, the syntax above shows an initial rule for defining nested determiners. This could support an expression like:

```
some of John's 5 accounts
(accounts (wusage noun) (number plural)
     (det
          (5
               (det
                    ('s (John (wusage noun) (naming proper))
                    (det (some
                              (of (wusage prep)]
```

The syntax above does not prevent generating nonsense nested determiners, such as "the of 5 her boxes". This is left as a topic for future work.

5.3.7   Adjectives

The Tala syntax for an adjective is:

```
<adjective> :=
   (adj {<adj-word>|<gerundive>|<c-conj-adj>}) |
   (adj (<adj-word>    ;a word defined in the Tala lexicon
                       ;as an adjective
        (wusage adj)
        <adjective>*
        <adverb>*
        <prep-link>*
        <pointer-bind>?
        (wsense <pointer>)? ;viz. §3.6.1
        (wreferent <pointer>)? ;viz. §3.6.1
        )

<gerundive> :=    ;past or present participle gerundive
   (<verb-word>  ;a word defined in the lexicon as a verb
        (wusage adj) ;here being used as an adjective,
```

```
                        ;in participle form
(tense {present | past})
(aspect {simple | continuous})?
(passive)?
(agreement-number singular)
(agreement-person non-I)
<adjective>*
<adverb>*
<prep-link>*
<pointer-bind>?
(wsense <pointer>)? ;viz. §3.6.1
(wreferent <pointer>)? ;viz. §3.6.1
)
```

Examples using the above syntax for adjectives:

> good in coffee
> old and new books
> eating place
> the eaten meal
> rapidly best selling item
> easily changed in a hurry costume

5.3.8   Adverbs

The Tala syntax for usage of an adverb is:

```
<adverb> :=
   (adv {<adv-word>|<c-conj-adv>}) |
   (adv
       <adv-word> ;a word defined in the Tala lexicon
                  ;as an adverb
      (wusage adv)
      <adverb>*
      <prep-link>*
      <verb>*
      <pointer-bind>?
      (wsense <pointer>)? ;viz. §3.6.1
      (wreferent <pointer>)? ;viz. §3.6.1
      )
```

This syntax allows concatenating adverbs (e.g. "very rapidly"), using conjunctions of adverbs (e.g. "quickly and efficiently") and modifying an adverb with a preposition ("he travelled rapidly by his method of measurement", "similarly to all purchases", "happily for now"…). 'Conjunctive adverbs' are used to relate multiple verbs: "The wind died, consequently we rowed the sailboat". Conjunctive adverbs could also be treated as subordinating conjunctions in the Tala syntax (§5.3.9.2).

5.3.9   Conjunctions

The Tala syntax for conjunctions is:

```
                <conjunction> := <c-conj> | <s-conj>
```

5.3.9.1 Coordinating Conjunctions

The Tala syntax for a coordinating conjunction is:

```
<c-conj> := <c-conj-noun>|<c-conj-verb>|<c-conj-adj>|
            <c-conj-adv>| <c-conj-prep>
<c-conj-word> := and | or
<c-conj-noun> := (<c-conj-word> (wusage conj)
                    <noun>+ <pointer-bind?>)
<c-conj-verb> := (<c-conj-word> (wusage conj)
                    <verb>+ <pointer-bind?>)
<c-conj-adj> :=
    (<c-conj-word> (wusage conj)
        <adjective>+ <pointer-bind?>)
<c-conj-adv> :=
    (<c-conj-word> (wusage conj)
        <adverb>+ <pointer-bind?>)
<c-conj-prep> :=
    (<c-conj-word> (wusage conj)
        <prep-link>+ <pointer-bind?>)
```

The following illustrates coordinating conjunctions with shared dependencies across multiple parents and dependents:

```
He and she buy and sell old and new cars and trucks cheaply
and profitably in Phoenix on Thursdays and Tucson on Fridays.

(and
    (wusage conj))
    (buy
        (wusage verb)
        (subj
            (and (wusage conj)
                 (<- ?subjs)
                 (he (wusage pronoun))
                 (she (wusage pronoun))))
        (obj
            (and (wusage conj)
                 (<- ?objs)
                 (car
                     (wusage noun)
                     (number plural)
                     (adj
                         (<- ?adjs)
                         (and (wusage conj)
                              (old (wusage adj))
                              (new (wusage adj))
                              )))
                 (truck
                     (wusage noun)
                     (number plural)
                     ?adjs)))
        (adv
            (<- ?advs)
            (and (wusage conj)
                 (cheaply (wusage adv))
                 (profitably (wusage adv))
```

```
                       ))
            (in
                (<- ?inprep)
                (wusage prep)
                (and
                        (wusage conj)
                        (Phoenix (wusage noun) (naming proper)
                            (on (Thursday wusage noun)
                                    (naming proper)
                                    (number plural))
                              ))
                        (Tucson (wusage noun)(naming proper))
                            (on (Friday wusage noun)
                                    (naming proper)
                                    (number plural))
                              )))))
        (sell (wusage verb)
            (subj ?subjs)
            (obj ?objs)
            ?advs ?inprep]
```

This represents a dependency grammar parsing of the example sentence. Conceptual processing would be responsible for creating separate interpretations if needed, in which "on Thursdays" modifies "buy and sell" in Phoenix and "on Fridays" modifies "buy and sell" in Tucson.

This example also illustrates the value of using pointers in Tala expressions, to maintain concision equivalent to English. The above example would be combinatorially larger if the use of Tala variables as pointers were not allowed. Using pointers, the Tala expression is isomorphic to a dependency parse tree for the English sentence, and in this sense it is as concise as English. This notion of equivalent concision could be expressed more precisely by saying that the Tala expression has size complexity $O(n)$, where n is the size of a dependency parse tree for an English sentence. Note that the pointers in the Tala expression correspond to extra arcs representing shared dependencies in a dependency parse diagram. [84] Per the structurality requirement to represent syntax in Tala (§3.5.2), this kind of equivalent concision is the best we can do: To make Tala expressions as short as linear strings of English words, in effect identical to English in concision, would sacrifice representation of syntactic structure.

---

[84] It is more precise to say a Tala expression is homomorphic to a dependency parse tree for an English sentence, since the inverse mapping from the parse tree to Tala would not restore information in (wsense) and (wreferent) expressions.

5.3.9.2 Subordinating / Structured Conjunctions

As noted in §5.3.4, some authors argue that most subordinating conjunctions should be treated as prepositions. The Tala syntax allows certain words to be used as either prepositions or subordinating conjunctions, much as some words may be used as either nouns or verbs. The reason for this is that certain words often used as subordinating conjunctions are important in representing executable concepts and, per §3.2.1, Tala has a theoretical requirement to provide the syntax of at least a simple programming language. Thus we could use "if" as a preposition in "one if by land", but in Tala we wish to use "if" in an "if-then-else" expression in an executable concept. In this case, "if" is no longer a preposition dependent on a verb. Nor is "if" even necessarily subordinate to a verb, since it may be at the outermost level of a Tala sentence. Hence for our purposes it is more accurate to refer to such conjunctions as "structured" rather than subordinating. (Arguably, the term 'conjunction' is not accurate either. It is kept since it is already accepted in linguistics for describing words that may be disjuncts.) The Tala syntax for structured conjunctions is:

```
<s-conj> :=
    <if-then-else-conj>|<how-conj>|<why-conj>|
    <when-conj>|<while-conj>|<until-conj>|<typical-s-conj>
```

If, Then, Else

This structured conjunction is used for conditional expressions in Tala executable concepts (xconcepts). The syntax is:

```
<if-then-else-conj> :=
  (if
      (wusage s-conj)
      <test>
      (then <verb>+)?
      (else <verb>+)?
      <pointer-bind>?
  )
<test> := <verb>|<c-conj-verb>
```

The `<test>` in this expression is evaluated by conceptual processing to determine if its verb expression exists in a context. If it does, then the `(then ...)` expression is processed, otherwise the `(else ...)` expression is. If multiple verbs occur inside a `(then ...)` or `(else ...)`, they are processed sequentially when an xconcept is interpreted. The s-conjunction `(steps ...)` may be used to list a sequence of verbs that can be performed sequentially, without specifying if-then-else. (This is further discussed in §5.5.)

How

This structured conjunction is used to refer to how something happens, and optionally to describe the method for performing an action.

```
<how-conj> :=
    (how
        (wusage s-conj)
        <verb>?
        (method <verb>)?
        <pointer-bind>?
        )

;example: how can I make grain be food for people?
(how
    (wusage s-conj)
    (make
        (wusage verb)
        (subj ?self)
        (modal can)
        (sentence-class question)
        (obj
            (be
                (wusage verb)
                (subj
                    (grain
                        (wusage noun)
                        ))
                (obj
                    (food
                        (wusage noun)
                        (for
                            (people
                                (wusage noun)
                                ]
```

Why

This structured conjunction is used to refer to why something happens, and optionally to describe the cause and/or purpose of an action.

```
<why-conj> :=
    (why
        (wusage s-conj)
        <verb>?
        (cause <verb>)?
        (purpose <verb>)?
        <pointer-bind>?)
```

When

This structured conjunction is used to refer to when a test is satisfied, and optionally to describe an action to perform at that time.

```
<when-conj> :=
    (when
```

```
(wusage s-conj)
<test>
(do <verb>)?
<pointer-bind>?
)
```

While

This structured conjunction is used to refer to the time period during which a test is satisfied, and optionally to describe an action to perform during this period.

```
<while-conj> :=
  (while
     (wusage s-conj)
     <test>
     (do <verb>)?
     <pointer-bind>?)
```

Until

This structured conjunction is used to refer to the time period before a test is satisfied, and optionally to describe an action to perform during this period.

```
<until-conj> :=
  (until
     (wusage s-conj)
     <test>
     (do <verb>)?
     <pointer-bind>?
     )
```

Other Subordinating Conjunctions

The syntax of other subordinating conjunctions would be consistent with representing them as prepositions (though Huddleston & Pullum reason that 'that' and 'whether' should remain treated as subordinating conjunctions):

```
<typical-s-conj> :=
  (<word>
   (wusage s-conj)
   <verb>
   <pointer-bind>?
   )

<word> := after | although | as | because | before |
          once | since | than | that | though |
          whether | so | which | ...
```

5.3.9.3   Correlative Conjunctions

Correlative conjunctions are words used together as conjunctions, e.g.:

> "both ... and"
> "either ... or"

"neither … nor"

"not only … but also"

Some of these could be treated as logically equivalent to coordinating conjunctions, while others could be represented as structured conjunctions. This is left as a topic for future work.

5.3.10   Interjections

Interjections are words that convey emotion, and may not be grammatically related to other parts of a sentence, e.g. "Wow, that's amazing!" It would be straightforward to extend the Tala syntax to allow interjections to be added to verbs. This is left as a topic for future work.

5.3.11   Tala Variables and Pointers

A Tala variable (`<tala-var>`) is any Scheme symbol that starts with "?" and has at least one other character, for example `?x`. Tala variables are untyped in the prototype. The Tala syntax for a Tala variable to be bound to a Tala expression is:

```
<pointer-bind> := ("<-" <tala-var>)
```

This syntax specifies that a Tala variable is bound to the expression containing the pointer-bind expression. Expressions like `(<- ?x)` occur throughout this thesis to illustrate use of pointers for concept representation in Tala.

A Tala variable may also occur in Tala expressions in place of any of the following: `<noun-word>`, `<verb-word>`, `<prep-word>`, `<pronoun>`, `<det>`, `<adj-word>`, `<adv-word>`. This enables pattern-matching logic to bind the variable to a corresponding word in a Tala expression. Section 5.5.3 provides further information on pattern-matching in the prototype.

5.3.12   Inflections

Hudson (2010) writes that English has only two rules of agreement for inflections: a rule specifying that a determiner agrees in number with its complement noun, and a rule specifying subject-verb agreement for number and tense. Per §3.4.1, the grammar for Tala is non-prescriptive: Tala allows sentences that have incorrect inflections because such sentences, though ungrammatical, are generated by people in the real world, and Tala should be able to represent how sentences are actually expressed. Even so, the Tala grammar should facilitate conceptual processing and representation of sentences that have grammatical inflections. The following pages describe how English rules of agreement for inflections can be supported in the Tala syntax.

5.3.12.1 Determiner-Complement Agreement

In the Tala syntax above, the number feature of a noun specifies whether its usage is singular or plural. The Tala lexicon could specify whether individual determiners indicate singular or plural number, as follows:

```
the - singular | plural ;(the box | the boxes)
a - singular ;(a box)
this - singular ;(this box)
these - plural ;(these boxes)
...
```

This information could be used for conceptual processing and representation of sentences that have internal agreement between pronouns and complement nouns. This is a topic for future research, which in principle is supportable in the TalaMind architecture.

5.3.12.2 Subject-Verb Agreement

The Tala syntax specified in previous sections includes Hudson's (1999) features for subjects and verbs, which support subject-verb agreement. The features for subjects (nouns and pronouns) are agreement-number and agreement-person. The features for verbs are subject-number and subject-person. Hudson's agreement rules may be restated as:

1. If a verb has a subject-number, then it must be the same as its subject's agreement-number.

2. For the verb BE if the subject is I, then the verb's subject-person and the subject's agreement-person must be the same.

The features number and agreement-number normally have the same value (i.e. singular or plural), but this default can be overridden:

I always has plural agreement-number and singular number.

Regardless of its number, the agreement-number for *You* is always plural.

However, Hudson notes that for some subjects (like *two drops* or *set*) the meaning of the subject allows these rules to be overridden.

Developing a formal account of this appears to be a topic for further study.

The feature `subject-number` applies only to present-tense non-modal verbs and BE. The form *was*, for example, is lexically defined as having singular `subject-number`. Other verbs, such as past-tense verbs, simply have no `subject-number`. The feature `subject-person` applies only to the one verb, BE, and has just one task: to distinguish forms that combine with I (i.e. *am* and *was*) from the other forms. Following is an example of these rules:

```
You were sneezing.
(be
    (wusage verb)
    (tense past)
    (subject-number plural)
        ;must agree with subject agreement-number
    (subject-person non-I) ;must agree with subject
        ;tense, subject-number and subject-person
        ;select 'were'
    (subj
        (you
            (wusage pronoun)
            (number singular)
            (agreement-number plural)
                ;always plural for you
            (agreement-person non-I)
            )
        )
    (obj
        (sneeze
            (wusage verb)
            (aspect continuous)
            ]
```

The TalaMind approach is not restricted to Hudson's rules for subject-verb agreement. Tala could accommodate other linguistic analyses than the ones presented here. In some thesis examples and prototype code, forms like `(subj-person third-singular)` are used instead, i.e. not following Hudson's approach to subject-verb agreement.

5.4 Design of Conceptual Framework

5.4.1 Requirements for a Conceptual Framework

Per §§3.2.2 and 3.6.7, following is a preliminary list of requirements for a TalaMind conceptual framework:

- Store concepts representing definitions of words.

- Store concepts representing natural language constructions.

- Store concepts representing current goals and thoughts.

- Store executable concepts, for conceptual processing.

- Represent semantic domains, mental spaces, conceptual blends, encyclopedic and commonsense knowledge.

- Represent a perceived / projected reality (§2.2.3) with perceptions of the environment (percepts) and actions in the environment (effepts).

- Represent an 'event-memory', storing knowledge of previous events.

- Represent interactive contexts and mutual knowledge.

- Represent theories, problem contexts, and composite contexts.

- Represent hypothetical contexts and support nested conceptual simulation.

The following sections describe how many of the above features have been implemented in the prototype TalaMind demonstration system, to a very limited extent necessary to support TalaMind simulations. Per §1.6, fully implementing all of them would be a long-term research effort, involving teams of researchers. More requirements for conceptual frameworks will be identified in future research on the TalaMind approach.

5.4.2 Structure of the Conceptual Framework

The prototype required a specific design for a conceptual framework, relative to the above requirements. Yet the TalaMind approach does not mandate any particular implementation technology, nor any particular design for a conceptual framework.

The above list may be viewed as a general requirement to manage Tala concepts in different locations [85] within a data structure (or collection of data structures) and to support different kinds of conceptual processing in different locations. For instance, one location could correspond to definitions of words, another to perceived reality,

---

[85] One could imagine a structure (such as a hologram or neural net) that did not put Tala concepts in different locations, if one wished to take a different approach.

another to event-memory, etc. Precisely how such locations are represented and accessed is a design choice.

When creating a prototype, it's necessary to start somewhere and "put a stake in the ground", recognizing that design choices may need to be revised later. So, it was important to choose a data structure that allows easily defining and representing locations for Tala concepts, and easily changing locations later, if needed. Also, per §3.5.2 the design choice in this thesis is to represent Tala concepts using Lisp list structures. A resulting design choice was to write the prototype in JScheme, a dialect of Lisp implemented in Java.[86] These design choices made list data structures natural options to consider for the conceptual framework. Finally, there was not a requirement for the demonstration system to support external corpora or large concept bases, and so there was not a requirement for efficiency and scalability in the prototype framework.

All these considerations led me to create a nested list of association lists, as the data structure for the prototype conceptual framework. This structure evolved during the prototype effort, as I decided somewhat arbitrarily where to store different kinds of Tala concepts within the framework. Following is its current state:

```
(mind
    (concepts
        (words)
        (xconcepts
            (percepts)
            (mpercepts)
            (goals)
            (constructs)
            ))
    (subagents general mu nu)
    (contexts
        (p-reality ;perceived / projected reality
            (percepts)
            (effepts)
            (mpercepts
                (general)
                (mu)
                (nu)
                )
            (meffepts
                (general)
                (mu)
                (nu)
```

---

[86] Besides supporting Lisp list structures, this enabled running the demonstration as an applet for thesis advisors to view on the Web.

```
            )
(constructs
    (general)
    (mu)
    (nu)
    )
(construct-buffer
    (general)
    (mu)
    (nu)
    )
(concepts
    (encyclopedia)
    (goals)
    (thoughts)
    (mental-spaces)
    )
(current-domains)
(event-memory)
(scenarios)
(systems)
]
```

It is not claimed that this list structure is optimal or required for future use – it is just part of a prototype. Tala concepts are inserted into and retrieved from the above list structure, and maintained from one time interval to the next, during a TalaMind simulation. The following pages discuss elements of this list structure in more detail.

There may be some advantages for use of list structures in future to represent portions of conceptual frameworks, for replication in nested conceptual processing: It was helpful in programming the prototype to have standard paths to concepts in different parts of the framework, which could be reused in nested conceptual simulation. Other portions of the framework, such as the Tala lexicon and encyclopedic knowledge, will need data structures such as hashtables to be scalable.

5.4.3   Perceived Reality – Percepts and Effepts

The (p-reality) slot stores a Tala agent's concepts for its perceived reality. It is located at the path (mind contexts p-reality) in the conceptual framework. Within perceived reality, Tala concepts representing perceptions and actions are stored in the (percepts) and (effepts) slots, respectively. The (percepts) slot is updated in each time interval to contain an agent's current perceptions. As a result of processing executable concepts an agent may update its (effepts) slot, indicating actions to be performed in the environment.

The prototype logic simulates a conceptual interface (§1.5, Figure 1-1) to the environment by sending effepts created by a Tala agent to other agents or external systems (e.g. the `grain` system), where they are processed as percepts in the perceived realities of other agents or as input-actions to simpler, behavioral systems (§5.4.17).

Since Tala concepts represent syntactic structure of natural language expressions, effepts can describe physical actions ("Ben pounds grain") or speech acts communicating concepts to other agents ("Ben says Leo try this flat bread"). Throughout this thesis, the term *speech act* is used according to Austin's (1962) description of a 'total speech act', which includes locutionary as well as pragmatic (illocutionary and perlocutionary) acts. Though some authors have used the term primarily referring to illocutionary acts, by default here it refers to locutionary acts, which may entail illocutionary or perlocutionary acts. Pragmatic aspects have been addressed by discussions of conceptual processing in contexts (§3.6.7), with abduction and encyclopedic knowledge.

5.4.4   Subagents, Mpercepts, and Meffepts

The prototype TalaMind system includes a society of mind (§2.3.3.2) for subagents within each Tala agent. These subagents can process executable concepts and communicate with each other by exchanging Tala concepts. Additionally, subagents can create external effepts for a Tala agent, and process the percepts of a Tala agent. The prototype provides three subagents for each Tala agent. These are called *mu, nu,* and *general*, prefixed by the name of the Tala agent, e.g. Ben-mu.

When a Tala subagent transmits a Tala concept for processing by other Tala subagents, it effectively performs a *mental speech act* (*"meffept"*). The concept is received by other Tala subagents as a mental percept (*"mpercept"*). The TalaMind logic treats mpercepts and meffepts analogously to percepts and effepts: meffepts are transmitted to other subagents and received by them as mpercepts, just as effepts are transmitted to other Tala agents and external systems and received by them as percepts or input-actions. Thus, the TalaMind prototype simulates self-talk (mental discourse) within a Tala agent.

5.4.5   Tala Lexicon

The `(words)` slot of the conceptual framework contains definitions of words, expressed in the Tala mentalese. It is located at the path `(mind concepts words)` in the conceptual framework. For example, following is a definition of food:

```
(food
     (wusage noun)
     (subj-of
          (means
               (obj
                    (object
                         (det any)
                         (obj-of
                              (eat
                                   (wusage verb)
                                   (modal can)
                                   (subj
                                        (animal
                                             (wusage noun)
                                             (det an)
                                        ]
```

This definition says "food means any object that an animal can eat". The definition could be improved, of course, but is adequate for the demonstration. For the prototype demonstration, several nouns and verbs are used effectively as primitive words, even though they could be given definitions in principle (cf. §3.6.8).

The TalaMind logic interprets some verbs to provide various kinds of processing within the conceptual framework. For example, "think" corresponds to a mental speech act by a Tala subagent, which may be perceived by other Tala subagents within the same Tala agent. Such speech acts are asserted and found in the (thoughts) slot of the conceptual framework. "Want" corresponds to asserting a concept in the (goals) slot of the conceptual framework.

5.4.6   Encyclopedic Knowledge and Semantic Domains

The (encyclopedia) slot stores encyclopedic knowledge for a Tala agent. For the prototype, encyclopedic knowledge has been created by hand, containing concepts in a few semantic domains. It uses the following structure for each domain:

```
(<domain-name> ;typically a noun, but could be a phrase
     (domain-matrix ;domains that this domain refers to
          )
     (concepts
          ;mentalese sentences in the domain
          ;includes a definition of the domain-name
          ))
```

This gives encyclopedic knowledge a structure similar to a conventional encyclopedia. For example, the prototype includes a semantic domain for knowledge about nuts:

```
(nut
     (domain-matrix plant)
```

```
(concepts
    a nut is an edible seed inside an inedible shell.
    to eat a nut a human must remove its
        inedible shell.
    humans can eat nuts removed from shells.
    humans can remove shells from nuts by pounding nuts
        because pounding breaks shells off nuts.
    ))
```

The prototype encyclopedic knowledge also includes small semantic domains for *grain* and *people*.

5.4.7   Current Domains

The `(current-domains)` slot stores Tala mentalese concepts representing current discourse domains for a Tala agent, to support semantic disambiguation. These concepts may be used as indexes into encyclopedic knowledge.

5.4.8   Mental Spaces and Conceptual Blends

The `(mental-spaces)` slot in the TalaMind conceptual framework holds mental spaces, which have the following structure:

```
(<space-number> ;unique # for this space
    (space-type ;e.g. "blend"
        )
    (elements ;Tala nouns in the space
        )
    (concepts ;Tala mentalese concepts in the space
        ))
```

This gives mental spaces a structure similar to semantic domains in encyclopedic knowledge, but allows mental spaces to be created separately and hold concepts developed tentatively (viz. §3.6.7.8). Conceptual blends are mental spaces that are created by blending concepts from other mental spaces (viz. §3.6.7.9).

5.4.9   Scenarios

The `(scenarios)` slot is used for creating nested concept frameworks that allow simulation of imagined future scenarios, i.e. nested conceptual simulation (viz. §6.3.5.2). These can be nested arbitrarily deep. Since scenarios are nested concept frameworks, they include much more structure than mental spaces, as implemented in the prototype: Scenarios include nested copies of projected reality, xconcepts, etc., to enable simulating a hypothetical reality inside a Tala agent's mind, including a Tala agent's simulation of other Tala agents' minds or its own mind in hypothetical situations.

5.4.10 Thoughts

Within the prototype, the verb "think" is used to refer to mental speech acts, and the demonstration can display sentences like "Ben thinks people need more food sources". The (thoughts) slot in the TalaMind conceptual framework holds Tala concepts developed and communicated as mental speech acts by subagents (§5.4.4). These concepts are communicated between subagents by the prototype's interpretation of the verb "think", which stores them in the subagent's (meffepts) slot and the agent's (thoughts) slot, and then copies them to the (mpercepts) slots of other subagents. These concepts may be retrieved via pattern-matching from the (thoughts) slot, using think as a pattern-matching verb within executable concepts.

Of course, the transmission of a thought does not determine what other thoughts it may engender, i.e. how other Tala subagents may process it and potentially respond with other thoughts. Likewise, interpreting the verb think to transmit a thought does not specify the conceptual processing that occurred in the creation of the thought – different forms of conceptual processing can occur to create Tala concepts before they are communicated by mental speech acts.

5.4.11 Goals

The (goals) slot stores the current goals of a Tala agent, within its perceived reality. In the TalaMind demonstration, a goal is a Tala conceptual structure using the verb "want". The object of a goal is itself a Tala mentalese expression. For example, a goal might be:

```
(want (wusage verb)
      (subj ?self)
      (obj
            (examine
                  (wusage verb)
                  (subj ?self)
                  (obj
                        (grain
                              (wusage noun)]
```

This goal says in effect "I want to examine grain", though it does not use an infinitive.

5.4.12 Executable Concepts

The (xconcepts) slot in the conceptual framework stores executable concepts. Within this slot, executable concepts are organized according to whether they match percepts, mpercepts, goals or support processing constructions. The design of executable concepts is discussed in §5.5.2.

5.4.13   Tala Constructions and Metaphors

The (constructs) slot stores Tala constructions (§3.6.3.13). The conceptual framework provides constructs and construct-buffer slots for processing concepts produced by using constructions, located at the following paths in the conceptual framework:

```
(mind contexts p-reality constructs)
(mind contexts p-reality construct-buffer)
```

These slots support translating Tala mentalese concepts to and from different forms. The slots have substructure to support processing of constructions by subagents of a Tala agent. The demonstration illustrates semantic disambiguation of metaphors by automatic translation via constructions (viz. §3.6.7.9).

5.4.14   Event-Memory

The (event-memory) slot stores a Tala agent's memory of percepts and effepts within perceived reality, for each timestep of a simulation. The prototype logic automatically updates this slot during a simulation, and supports a Tala agent searching its event-memory slot to recall events. Note: At present the prototype does not store a Tala agent's memory of its subagent's mpercepts and meffepts, nor trace information about the executable concepts that produced them. These could be added in future research, to study reflection about thought processes.

5.4.15   Systems

The (systems) slot stores a Tala agent's concepts about other systems (agents and objects) within its perceived reality. Its main use in the prototype is to store an agent's internal name for grain, so that it can be referred to as flour, dough, flatbread, or leavened bread, as it is gradually transformed during the demonstration.

5.4.16   The Reserved Variable ?self

For each Tala agent, the reserved variable ?self is automatically bound to an internally unique identifier for itself when the Tala agent is created. For the prototype simulations, this unique identifier is just the name of each Tala agent, i.e. Ben or Leo. Conceptual processing is responsible for translating Tala concepts using first-person singular pronouns into Tala concepts with the appropriate binding. For example, if Ben perceives that Leo says "I want X", then Ben's conceptual processing should translate this into "Leo wants X" (or whatever internal identifier or pointer value that Ben uses to refer to Leo). Constructions may be used to perform these translations (§5.5.4).

5.4.17   Virtual Environment

For the TalaMind prototype, a virtual environment is needed in which Tala agents can communicate with each other and interact with other, simpler systems representing objects in the environment. My strategy in creating the TalaMind demonstration was to minimize work on implementing a virtual environment, and to avoid simulating spatial perceptions, graphics display, etc. So, for the demonstration a virtual environment for Tala agents is represented very simply by a list structure, called the `reality` list structure. It contains a collection of list structures representing systems, which contain information for Tala agents and simpler behavioral systems such as objects that can change state:

```
(reality
    (systems
        (<system-id> ;a unique id for each system,
                     ;e.g. Ben or grain
            (body)
            (behavior
                (input-actions)
                (output-actions)
                (transitions) ;xconcepts
                (current-state)
                (start-state)
                (end-states))
            (mind
                 . . .
            ]
```

The information for each system in the TalaMind `reality` list structure has potentially three substructures, called *body*, *behavior*, and *mind*:

- The (mind) substructure is the conceptual framework of a Tala agent described in the preceding sections, §5.4.1 through §5.4.16. It stores Tala mentalese conceptual structures representing the mind of a Tala agent, i.e. an agent's concepts about the environment, other agents, and itself.

- The (behavior) substructure of a system supports storing and representing the behavior of a system expressed as a finite-state machine, using Tala mentalese conceptual expressions. So, the (behavior) substructure supports modeling systems external to Tala agents, which do not have the internal complexity of

Tala agents.

In the demonstration, this is used to represent the behavior of wheat grains as Ben performs actions that eventually transform wheat into bread, and Leo tastes the results. The slots for – `state` information contain Tala mentalese expressions describing states of a system. The (`transitions`) slot contains Tala executable concepts for changing state and/or performing `output-actions`, in response to `input-actions` on a system. Both output and input actions are also represented as Tala mentalese expressions. So a system with only a (`behavior`) substructure is easily described and interfaced with systems that have (`mind`) substructures. During each timestep, the prototype processes interactions between the conceptual frameworks of Tala agents, and behaviors of simpler systems.

- The (`body`) substructure of a system is a stub for future use, to represent the physical structure of a system in a form that could support display and simulation, separate from whatever is represented within a system's (`mind`) or (`behavior`). Perhaps (`body`) may eventually have a structure similar to a computer graphics scenegraph, to allow description and display of a hierarchical structure of three-dimensional graphics objects. This could eventually support a three-dimensional, graphical display of TalaMind systems interacting in a simulation.

5.5 Design of Conceptual Processes

5.5.1 TalaMind Control Flow

Routines written in JScheme implement a control flow for conceptual processing in the TalaMind demonstration system. In each timestep of a simulation, the prototype logic performs conceptual processing for both Tala agents and their subagents, and for behaviors of systems external to Tala agents, by calling the following functions written in JScheme:

```
processAgentEffepts
    process the effepts generated by each Tala agent, causing
    input actions to Tala behavioral systems, and generating
    percepts by other Tala agents.

processSystemsOutputActions
    process the output actions generated by Tala behavioral
    systems, generating percepts for Tala agents (and
    potentially, input actions for behaviors of other TalaMind
    systems).
```

```
processSystemsBehaviors
    for each Tala behavioral system, process its behavior
    methods for input-actions it has received during this
    timestep, to change state, generate output actions, and
    create percepts for Tala agents.

processAgentsTalaMinds
    for each Tala agent, perform processing of its (mind)
    substructure.
```

In the function `processAgentsTalaMinds`, processing of a Tala agent's `(mind)` substructure consists of iterating in a loop[87,88] executing the following functions (written in JScheme) until no new Tala conceptual structures are created by interpretation of executable concepts written in Tala:

```
processAgentPercepts
processAgentMeffepts
processAgentMpercepts
updateAgentConstructs
processAgentConstructs
processAgentGoals
processAgentProcesses
removeAgentPercepts[89]
removeProcessedGoals
```

Following are pseudocode descriptions of the above functions:

```
processAgentPercepts
    for each percept in (contexts reality percepts)
        for each subagent in (mind subagents)
            for each xconcept in
                (concepts xconcepts percepts subagent)
                if the xconcept matches the percept
                then perform the xconcept

processAgentMeffepts
    for each subagent in (mind subagents)
        for each meffept of the subagent
            display flatenglish or concept structure
            of the meffept
            copy the meffept to mpercepts of all
            other subagents
        delete meffepts of the subagent
```

---

[87] In a simulation, agents generate output in different timesteps because each agent is driven by percepts of events generated by the other agent in a previous timestep.

[88] Executable concepts may generate new effepts, new meffepts, or new constructs in each loop iteration.

[89] This is a no-op after the first time through the loop, since an agent does not receive new percepts until the start of a timestep.

```
processAgentMpercepts
    for each subagent in (mind subagents)
        for each mpercept of the subagent
            for each xconcept in
                (concepts xconcepts mpercepts subagent)
                    if the xconcept matches the mpercept
                    then perform the xconcept
        delete mpercepts of the subagent

updateAgentConstructs
    for each subagent in (mind subagents)
        move all concepts
            from (mind contexts p-reality
                    construct-buffer subagent)
            to (mind contexts p-reality
                    constructs subagent)

processAgentConstructs
    for each subagent in (mind subagents)
        for each concept in
          (mind contexts p-reality constructs subagent)
            for each xconcept in
              (mind concepts xconcepts
               constructs subagent)
                if the xconcept matches the construct
                then perform the xconcept
        delete concepts from (mind contexts p-reality
                                constructs subagent)

processAgentProcesses
    for each process saved from a previous time-interval
        if its do-condition and wait-condition
                are satisfied
            then resume the process after the point
                it was suspended

processAgentGoals
    for each goal in (concepts goals) not already
      in-process or processed
        determine if the goal has been satisfied
                (i.e. if its object exists where specified
                by the goal, where this is determined
                by pattern matching candidate
                concepts with goal-obj)
        if the goal has been satisfied, then mark it
            as processed
            (so that it will be automatically removed)
        else
            if the object of the goal is to
              know-definition of something
                    (i.e. to pattern-match a
                    (know-definition ...) concept at
                    (concepts thoughts))
            then
                    call (getDefinition goal-obj ...)
                    ProcessXconceptsForGoal goal)
                    Mark the goal as processed
```

```
                  else call (ProcessXconceptsForGoal goal)

       getDefinition goal-obj
           (goal-obj is of form (know-definition ... (obj ...)),
           where the obj of know-definition is a Tala
           concept sentence.)

           if the goal-obj has not previously checked
               then call (hasTalaDefinitions?
                           know-definition-obj ...)
                       this will create a list of undefined words in
                       know-definition-obj

                       for each undefined word, create a new goal,
                           to have a definition for the word
                           (the agent has an xconcept to ask for
                           word definitions, to satisfy these goals)

       ProcessXconceptsForGoal goal
           for each xconcept at (concepts xconcepts)
               if the xconcept subject matches the goal
                   then interpret the xconcept
                       (assert the xconcept's object)
                       (this may cause a new goal or
                       a new xconcept to be created)
```

In the prototype, the above control flow is performed by a single-threaded, sequential JScheme process running in Java. Thus, Tala agents are processed sequentially in each timestep, and Tala subagents are processed sequentially for each Tala agent. As the structure of the prototype conceptual framework indicates (§5.4.2), a Tala agent's subagents each have access to all its executable concepts, percepts, goals, encyclopedic knowledge, etc. Tala subagents have individual mpercepts, meffepts, and concepts produced by using constructions (§§5.4.13, 5.5.4). In a more full and scalable design of a TalaMind architecture, it would be natural to use multi-threaded, parallel processing, and for Tala subagents to have other individual concepts and xconcepts, supporting a society of mind (§2.3.3.2). These are topics intended for future research.

5.5.2 Design of Executable Concepts

Extending the discussion in §3.2.1, we can identify several requirements for the design of executable concepts relative to the TalaMind architecture for the prototype: An executable concept should be able to describe steps to perform that may include external actions and speech acts (effepts), internal mental actions and speech acts (meffepts), and assertions or deletions of concepts in the conceptual framework, including creation and modification of other executable concepts. An executable concept should be able to specify conditions that may include tests on percepts, goals, finding concepts within the conceptual framework, etc. The design of executable concepts for the demonstration system provides these features via the following design elements:

- Use of structured conjunctions (if, then, else, steps, while, until) to provide syntax for conditional and iterative control expressions in Tala executable concepts.

- Use of do and wait to specify an executable concept should operate or pause and resume across time intervals of a TalaMind simulation.

- Use of the structured conjunction how to support defining an executable concept as the method to be executed in performing a verb.

- Use of and, or, not to support specifying and testing conditions.

- Use of pattern-matching logic to locate and bind variables to (parts of) concepts in the conceptual framework based on partial specifications.

- Use of the verb insert-step with prepositions into and after to support modifying executable concepts, by adding a new step before or after other steps in the xconcept.

- Use of verbs think, say, want to support creation of internal and external speech acts, and goals, respectively, and to support matching previously created / perceived speech acts and goals, within the conceptual framework.[90]

For example, during part of the discovery of bread simulation a Tala executable concept with the following pseudocode description is processed across multiple time intervals:

---

[90] say, think, and want are verbs interpreted specifically in the design of the prototype, for support of speech acts (effepts), mental speech acts (meffepts), and goals. Other verbs (e.g. decide, infer, or plan) could have specific interpretations in future designs of TalaMind architecture.

```
do
    steps
        think how can I make ?x softer
        random-affect ?x
        examine ?x
        wait until perceive ?x
        try to eat ?x
        wait until perceive ?x
    until
        or
            I think ?x is soft
            and
                I think ?x is a ?adjective ?substance
                I think ?adjective means soft
            I think ?x is a ruined mess
```

The prototype TalaMind logic supports defining an executable concept as the 'how' method of a verb. When the verb `random-affect` is processed its definition is looked up in the lexicon, and a Tala xconcept with the following pseudocode description is found and executed:

```
random-affect ?x
    means
        affect ?x
            adv randomly
            how
                method
                    random-xor-execute
                        mash ?x
                        pound ?x
                        soak ?x in water
                        mix ?x in water
```

The Tala primitive verb `random-xor-execute` randomly chooses one of the verbs within its scope, and executes it. Thus, `random-affect` results in a random action on ?x. In the demo, ?x is bound to `grain`. The prototype TalaMind logic transfers the random action (effept) to the finite-state behavior model of `grain`, which may cause `grain` to change state. The introduction of a random-choice verb provides a basic feature needed for demonstrations that do not follow a single, predefined flow.

If a verb does not have a 'how' method defined in the lexicon, and it is not predefined as a Tala primitive verb, then it is treated by default as an effept, i.e. an action to be transmitted to the object of the verb. Thus `mash`, `soak`, `pound`, `mix` are treated as effepts by default, and transmitted as input actions to the finite-state behavior model of `grain`.

Tala may be considered a universal programming language. To be universal, a programming language need provide only three basic control structure mechanisms:

- *Sequential* execution of one statement followed by another.
- *Conditional* execution of one statement or another, based on the value of a Boolean variable.
- *Iterative* execution of statements, until a Boolean variable is true.

This result is generally attributed to Bohm and Jacopini (1966); see Harel (1980) for a detailed history, and Fleck (2001, pp.115-119) for a concise treatment, based on Knuth and Floyd (1971).

The ability to define an executable concept as the 'how' method of a verb provides a means of abstraction in Tala, allowing complex behaviors to be named and accessed as units. Tala provides operations to create and modify concepts within the conceptual framework, using variables bound by pattern-matching.

Since executable concepts are written in the same syntax as other Tala concepts, they can pattern-match and process other executable concepts, in addition to ordinary, non-executable Tala concepts.

An xconcept can operate across multiple levels of the conceptual framework for a Tala agent. For example, an xconcept might match a concept in (contexts reality percepts) while the method might match or assert concepts in (contexts reality concepts thoughts) or in a nested scenario context.

Xconcept-based inference (using if-then-else expressions) can operate directly on natural language–based conceptual structures, without use of other logical formalisms.

5.5.3   Pattern-Matching

Per §3.6.6, Tala's use of nested list structures to represent syntactic structures facilitates pattern-matching of Tala sentences, which, combined with a syntax for variables in Tala and representation of inference rules as if-then sentences, enables the mechanics of logical inference. For the prototype demonstration system, I designed and implemented pattern-matching logic in JScheme so that a Tala variable can match anything in a multi-level Tala conceptual structure. Variable bindings are then used to instantiate the conceptual structures asserted by Tala executable concepts. The Tala pattern matcher is flexible in matching concepts even if attributes (slots) are specified in a different order from a pattern, and will match a concept that has more attributes than those specified by a pattern.

The demonstration system supports an unlimited number of Tala variables, which are created and bound dynamically only as needed for each Tala executable concept. A Java hashtable supports accessing variable bindings associated with each process instance of a Tala executable concept. This supports an executable concept running as a process across multiple Tala time intervals.

The system also supports binding Tala variables to the sequence numbers of nested contexts, so that these can be referenced in Tala concepts. This enables the following output in the farmer's dilemma simulation:

```
1...2    Leo thinks scenario 1.1 is a win-win.
```

For the demonstration system, this author's priority in writing the pattern matcher has been to achieve flexibility very quickly, to match arbitrarily nested conceptual structures, without concern for efficiency of pattern-matching. Improving efficiency is a topic for future research. At present it is not a problem that limits human observation of the simulations.

5.5.4   Tala Constructions

Tala constructions (viz. §3.6.3.13) are implemented as executable concepts in the TalaMind demonstration system using a primitive verb subformtrans. An example occurs in the discovery of bread story, where Ben generates an internal speech act (meffept) that the system displays as:

```
Ben thinks humans perhaps must remove shells from grains to
eat grains.
```

In this sentence, the preposition "to" indicates a temporal precedence relationship, between removing shells and eating nuts. This is translated using the construction:

```
(subformtrans
    (wusage verb)
    (subj
        (?do1
            (wusage verb)
            (modal must)
            (subj ?c-subj)
            (prep
                (to
                    (?do2
                        (wusage verb)
                        )))))
    (obj
        ;do1 must precede do2
        (precede
```

```
(wusage verb)
(modal must)
(subj
      (?do1
            (wusage verb)
            (subj ?c-subj)
            )
      )
(obj
      (?do2
            (wusage verb)
            (subj ?c-subj)
            ]
```

The subject pattern of this construction matches Ben's sentence (binding ?do1 to "remove", ?do2 to "eat", and ?c-subj to "humans") and produces the object pattern as the translation. In doing this, it automatically copies information not specified in the subject pattern (e.g. "shells", "grains", and the prepositional phrase "from grains") into appropriate locations for instantiation of the object pattern. So, these may be considered as "underspecified constructions".

The conventions followed for automatic copying are:

- Information found in a concept matched by a construction's subject but not specified in the construction's subject, is copied to the construction's object by default.

- Information specified in the subject of a construction is kept in the construction's object only if specified in the construction's object.

A Tala grammatical construction can be matched at any location within a Tala concept. Thus, Ben's sentence above has an outer level "Ben thinks ...". The construction pattern-matching recurses within Ben's sentence to match "humans must remove shells ...", and performs replacement within the construct generated, at the location where the construction was matched. So, the output generated by the construction is:

```
(Ben translates as) Ben thinks humans remove shells
 from nuts must precede humans eat nuts.
```

Hence, Tala grammatical constructions are composable: Multiple constructions can perform translations in combination on different parts of a Tala concept.

The subformtrans construction produces output concepts that go into a Tala subagent's construct-buffer, which may be processed by other xconcepts or constructions. I also implemented a construction called `subformtransthink`, which uses the same logic to produce meffepts (mental actions) for a Tala subagent. This can support generation of concise metaphorical expressions as internal speech acts, which xconcepts can output as physical speech acts to other Tala agents.

The demonstration system's representation and processing of Tala grammatical constructions is fairly general in some respects, though more work remains for the future. As with semantic disambiguation in general, the goal has been to show one way that constructions could be supported within the TalaMind approach, without claiming a completely general method or solution at this point.

The generality of constructions is based on several characteristics:

- Both the subject and object of a construction may be arbitrarily simple or complex Tala mentalese conceptual structure patterns.

- Tala variables may appear anywhere within a construction subject or object, and be bound to arbitrarily simple or complex structures.

- A construction's subject pattern can be matched anywhere within a Tala mentalese expression, and replaced at that location by the instantiated object pattern.

- Constructions are composable: They may be used in combination with each other to transform different subforms of a Tala mentalese expression.

- Constructions are underspecified: Arbitrarily simple or complex conceptual structures not specified in a construction are processed according to a simple convention.

This generality is illustrated by how constructions are used in several steps of the demo:

> To translate "turn X into Y" as meaning "make X be Y".
>
> To translate "turn over X to Y" as meaning "give X to Y".
>
> To translate "must X to Y" as meaning "X must precede Y".
>
> To translate "kick in X" as meaning "give X to me".

To translate "me" into a reference to the agent making an utterance.

To translate "you" into a reference to the agent perceiving an utterance.

However, these examples also indicate some important limitations in how Tala constructions are currently implemented. In several cases there are exceptions when these constructions do not apply, and mechanisms are not yet implemented to handle such exceptions. Thus, "turn X into Y" may be used in situations that are not metaphorical, where "turn" refers literally to a change of physical orientation:

"turn the car into the driveway"

"turn the boat into the wind"

We would not wish[91] to translate these instances as:

"make the car be the driveway"

"make the boat be the wind"

One way to implement such conditionality would be to support processing of "suchthat" expressions, which would restrict application of constructions, along the following lines:

```
(subformtrans
      (wusage verb)
      (subj
            (turn
                  (wusage verb)
                  (subj (?s (wusage noun)))
                  (obj (?x (wusage noun)))
                  (into (?y (wusage noun)))
                  (suchthat
                        ;?x does not normally change
                        ;physical orientation relative
                        ;to ?y
                        )))
      (obj
            (make
                  (wusage verb)
                  (subj (?s (wusage noun)))
                  (obj
                        (be
                              (wusage verb)
                              (subj (?x (wusage noun)))))
```

---

[91] Though such metaphorical mistakes could be interesting as a source of new ideas...

```
(obj (?y (wusage noun)))
]
```

I did not implement "suchthat", because the present storyline does not cover enough different semantic domains to show these conditions positively: Implementing "suchthat" would not change the current "discovery of bread" storyline.

Similar limitations of current Tala construction logic are apparent in comparison to discussions given by other researchers. Constructions could be integrated with nested conceptual simulation, to support semantic disambiguation -- viz. Tyler and Evans' (2003) discussion of semantic disambiguation for the preposition "over". More work could be done to demonstrate the "evokes" characteristic of Embodied Construction Grammar, i.e. the ability of a word or concept to activate related concepts (cf. Feldman, 2006, p.289).

It appears the most general way to implement semantic disambiguation for constructions would be to fully implement conceptual blends (§3.6.7.9). A partial implementation of conceptual blends in the TalaMind framework is described later.

In principle such limitations in `subformtrans` may also be addressed in more general executable concepts, which can have arbitrarily complex logic, use nested conceptual simulation, and assert multiple concepts into the conceptual framework. Since Tala xconcepts match and assert natural language–based conceptual structures, and since Tala uses natural language–based conceptual structures to represent semantics, it follows that Tala executable concepts (xconcepts) are also in effect cognitive linguistics 'constructions': pairings of natural language forms with natural language semantics.

5.5.5   Tala Processing of Goals

The TalaMind prototype logic supports dynamic creation of goals and pattern-matching of goals with xconcepts, and automatically detects when goals have been satisfied. The logic also automatically deletes goals that have been satisfied and prevents attempting to process goals that are already being processed by other xconcepts. However the logic does not automatically propagate satisfaction between goals, nor automatically retract goals if they are no longer needed. Per §1.6, this functionality was not implemented in the prototype, since it has been well studied in previous research. It could be an enhancement in a future version of the system.

5.6   Design of User Interface

The TalaMind prototype demonstration was written using JScheme in a Java applet, so that simulations could be run by thesis reviewers using a Web browser that supports Java.

5.6.1   Design of the TalaMind Applet

When the user initially views the TalaMind applet, the panel shown in Figure 5-1 is displayed in the web browser. The center area is a text display field. During a TalaMind simulation, text is displayed showing concepts developed and communicated by Tala agents, and events that happen in the TalaMind environment. These are displayed as English sentences, with an option to view them as Tala concept list structures.

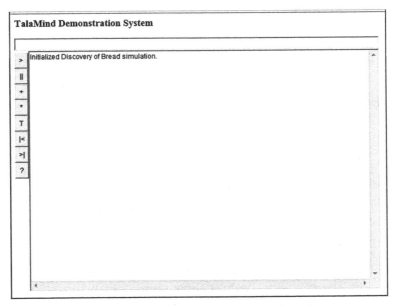

Figure 5-1 Initial Display of TalaMind Applet

Following is output produced by running the TalaMind simulation:

223

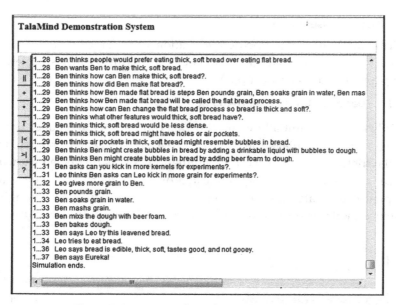

Figure 5-2 Output of a TalaMind Simulation

The buttons on the left side of the applet have the following functions:

>   run simulation

| | pause simulation

+   step simulation one time interval

| <   return to start of simulation, or go to previous simulation

> | advance to next simulation

*   toggle output between English and Tala concept display

T   display Tala concept framework

?   displays the above help information.

By default, the TalaMind applet runs the discovery of bread simulation. The | < and > | buttons allow running earlier versions of the simulation, and the farmer's dilemma simulation.

Using * to toggle the output to show Tala concepts will produce the following output for a simulation (Figure 5-3):

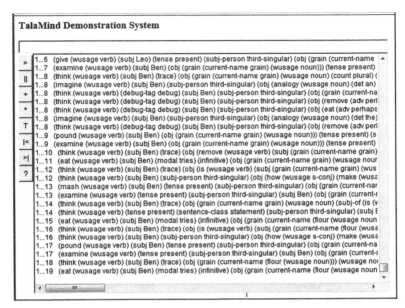

Figure 5-3 Tala Concepts Created During a Simulation

The T button displays concepts in the TalaMind framework. Its default setting is to display all percept xconcepts for Ben:

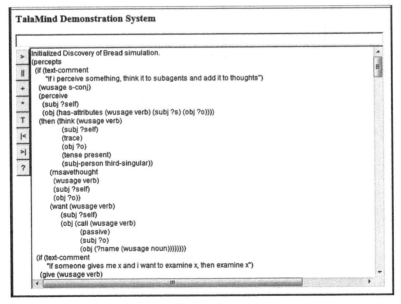

Figure 5-4 Display of Ben's Percept Xconcepts

225

In addition to printing information about the other buttons, the help button ? also prints information about TalaMind applet commands:

(sa)	show subagents in mental speech acts
(sc)	show construction processing
(sx)	prettyprint xconcepts when processed
(dl n)	set debug level to n (default is 0)
(show '(*path*))	show Tala concept framework at *path*

Just above the center area of the applet is a text input field, for optional use to enter these commands. It supports evaluation of Scheme expressions entered by hand, for which evaluation results are displayed in the text display field. If the user enters (show '(reality)) in the text input field, the system will pretty-print all the concepts for Tala agents in the TalaMind environment. The path specified in a (show ...) command will then be reused by the T button.

If the user types both the (sa) and (sc) commands, then the system will display output looking like this:

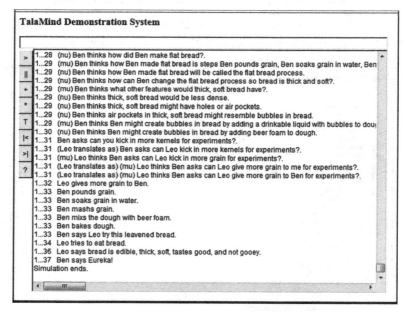

Figure 5-5 Display of Subagent and Construction Processing

The (sa) command displays the names of subagents (mu, nu, and general) producing internal speech acts within the storyline.

Using the (sc) command, additional steps will be displayed with prefixes of the form (Ben translates as) to show construction processing of Tala mentalese expressions. The source expression may be either a percept or mental percept (internal speech act), or a construct from a previous translation. Constructions operate in combination to perform different kinds of disambiguation. For example, some constructions translate metaphorical expressions, others translate personal pronouns (disambiguating "you" and "me" in separate steps), etc.

If the user types (sx) the system will prettyprint xconcepts when they are executed during the demonstration, as shown in Figure 5-6.

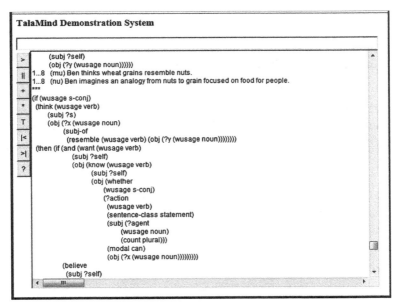

Figure 5-6 Display of Xconcept Execution During a Simulation

5.6.2   FlatEnglish Display

To make a TalaMind simulation easily understandable to people, the demonstration system uses logic (called FlatEnglish) written in JScheme to translate Tala conceptual structures into English sentences for display in the user interface. For example, the Tala conceptual structure:

```
(ask
   (wusage verb) (tense present)
   (subj Leo)
   (indirect-obj Ben)
   (obj
        (turn
              (wusage verb)
              (modal can)
              (sentence-class question)
              (subj you)
              (obj
                   (grain
                        (current-name grain)
                        (wusage noun)
                        ))
              (into
                   (fare (wusage noun)
                        (for
                             (people (wusage noun))
                             )))))
   (subj-person third-singular]
```

is translated by FlatEnglish logic into the text string:

```
Leo asks Ben can you turn grain into fare for people?
```

FlatEnglish is used throughout the demonstration to display speech acts between Tala agents, as well as to display actions performed by agents, internal speech acts by agents, etc. However, all communication between Tala agents (and within a Tala agent) is actually performed using Tala concept structures.

FlatEnglish may be considered as a demonstration that in principle Tala concept structures can be translated into written text and used for speech generation. Its use enables the TalaMind demonstration to focus on concept processing, rather than natural language parsing or generation.

5.7 Summary

This chapter presented a design for a prototype demonstration system, in accordance with the analysis of Chapter 3. The design for the syntax of the Tala conceptual language is fairly general and flexible, addressing issues such as compound nouns, gerunds, compound verbs, verb tense, aspect and voice, nested prepositions, clitic possessive determiners, gerundive adjectives, shared dependencies, coordinating and subordinating / structured conjunctions, subject-verb agreement, etc. This coverage indicates a Tala syntax could be comprehensive for English, though developing such a comprehensive syntax is a large effort that could occupy multiple researchers.

The design for a prototype conceptual framework includes representations of perceived reality, subagents, a Tala lexicon, encyclopedic knowledge, mental spaces and conceptual blends, scenarios for nested conceptual simulation, executable concepts, grammatical constructions, and event memory. The design for prototype conceptual processes includes interpretation of executable concepts with pattern-matching, variable binding, conditional and iterative expressions, transmission of mental speech acts between subagents, nested conceptual simulation, conceptual blending, and composable interpretation of grammatical constructions.

6. Demonstration

By consequence, or train of thoughts, I understand that succession of one thought to another which is called, to distinguish it from discourse in words, mental discourse. When a man thinketh on anything whatsoever, his next thought after is not altogether so casual as it seems to be. Not every thought to every thought succeeds indifferently...In sum, the discourse of the mind, when it is governed by design, is nothing but seeking, or the faculty of invention...a hunting out of the causes, of some effect, present or past; or of the effects, of some present or past cause...

~ Thomas Hobbes, *Leviathan*, 1651[92]

∞

6.1 Overview

The previous chapter presented the design of the TalaMind prototype demonstration system. This chapter discusses the execution of the demonstration system. It first presents the content of the simulations produced by running the system, and then discusses how these simulations illustrate that the TalaMind approach can potentially support the higher-level mentalities of human-level intelligence. Appendix B gives a step-by-step description of processing within the system for one of the demonstration simulations.

As noted previously, per §1.6 the prototype system cannot be claimed to actually achieve the higher-level mentalities of human-level intelligence, it can only illustrate how they may eventually be achieved. This use of illustration as a limited form of demonstration is similar to how a drawing may illustrate a fully operational machine. A drawing may outline the shape and structure of a machine and describe its operation without providing all the details needed to construct the machine. In the same way, the prototype simulations use functioning code for conceptual processing to show how higher-level mentalities

---

[92] As quoted by William James in *The Principles of Psychology*, I, pp.395-396, 1890. Hobbes' original spelling and capitalization were very different, e.g. 'train' was spelled 'trayne'.

could potentially be supported, though without encyclopedic and commonsense knowledge in a scalable, efficient architecture. These are needed to achieve human-level AI and are topics for the future, to leverage research in areas previously studied by others (viz. §§1.6, 7.6).

6.2 Demonstration Content

Per §5.2, the demonstration system is a functional prototype in which two Tala agents, named Ben and Leo, interact in a simulated environment. Each Tala agent has its own TalaMind conceptual framework and conceptual processes. To the human observer, a simulation is displayed as a sequence of English sentences, in effect a story, describing interactions between Ben and Leo, their actions and percepts in the environment, and their thoughts. The story that is simulated depends on the initial concepts that Ben and Leo have, their initial percepts of the simulated environment, and how their executable concepts process their perceptions to generate goals and actions, leading to further perceptions and actions at subsequent steps of the story. For the thesis demonstration, two stories have been simulated in which Ben is a cook and Leo is a farmer. The first is a story in which Ben and Leo discover how to make bread. The second is a story in which Ben and Leo agree to an exchange of wheat for bread, and then perform the exchange.

6.2.1 The Discovery of Bread Story Simulation

Initially in this story, neither Ben nor Leo knows how to make bread, nor even what bread is, nor that such a thing as bread exists. We may imagine Leo is an ancient farmer who raises goats and grows wheat grasses for the goats to eat, but does not yet know how to eat wheat himself. Ben is an ancient food and drink maker, who knows about cooking meat and making beer, presumably from fermented wheat grass.

The discovery of bread simulation includes output from a pseudorandom 'discovery loop': After removing shells from grain Ben performs a random sequence of actions to make grain softer for eating. This eventually results either in the discovery of dough, or in making grain a "ruined mess". In the first case, Ben proceeds to discover how to make flat bread, and then leavened bread. In the second case, he says the problem is too difficult, and gives up. Following is an example of output for the first case:

```
Initialized Discovery of Bread simulation.
1...1   Leo has excess grain.
1...1   Leo thinks Leo has excess grain.
1...2   Leo tries to eat grain.
1...3   Leo wants Ben to make edible grain.
1...4   Leo says grain is not edible because grain is too hard.
1...4   Leo asks Ben can you turn grain into fare for people?.
1...4   Ben thinks Leo says grain is not food for people.
1...4   Ben thinks Leo asks can Ben turn grain into food
        for people?.
1...4   Ben thinks why should Ben make grain be food for
        people?.
1...4   Ben thinks people need more food sources.
1...4   Ben thinks if Ben feasibly can make grain be food for
        people then Ben should make grain be food for people.
1...4   Ben wants Ben to know whether humans perhaps can
        eat grain.
1...4   Ben wants Ben to know how Ben can make grain be food
        for people.
1...4   Ben wants Ben to experiment with grain.
1...4   Ben wants Ben to examine grain.
1...5   Ben asks can you turn over some to me for experiments?.
1...5   Leo thinks Ben asks can Leo turn over some grain to
        me for experiments?.
1...5   Leo thinks Ben asks can Leo turn over some grain to
        Ben for experiments?.
1...6   Leo gives some grain to Ben.
1...7   Ben examines grain.
1...8   Ben thinks wheat grains resemble nuts.
1...8   Ben imagines an analogy from nuts to grain focused
        on food for people.
1...8   Ben thinks grain perhaps is an edible seed inside
        an inedible shell.
1...8   Ben thinks humans perhaps must remove shells from
        grains to eat grains.
1...8   Ben thinks humans perhaps can eat grains removed
        from shells.
1...8   Ben imagines the analogy from nuts to grain focused
        on removeing shells.
1...8   Ben thinks humans perhaps can remove shells from
        grains by pounding grains because pounding
        breaks shells off grains.
1...9   Ben pounds grain.
1...9   Ben examines grain.
1...10  Ben thinks grain is removed from shells.
1...11  Ben tries to eat grain.
1...12  Ben thinks grain is not edible because grain
        is very hard.
1...12  Ben thinks how can Ben make softer grain?.
1...13  Ben soaks grain in water.
1...13  Ben examines grain.
1...14  Ben thinks unshelled grain is soaked in water.
1...15  Ben tries to eat grain.
1...16  Ben thinks grain is not edible because grain is
        rather hard.
1...16  Ben thinks how can Ben make softer grain?.
```

```
1...17   Ben mashs grain.
1...17   Ben examines grain.
1...18   Ben thinks grain is a gooey paste.
1...18   Ben thinks grain that is a gooey paste will
         be called dough.
1...19   Ben tries to eat dough.
1...20   Ben thinks dough is soft, too gooey, and tastes bland.
1...20   Ben thinks dough is too gooey.
1...21   Ben bakes dough.
1...21   Ben examines baked dough.
1...22   Ben thinks baked dough is a flat, semi-rigid object.
1...22   Ben thinks baked dough that is a flat object will
         be called flat bread.
1...23   Ben tries to eat flat bread.
1...23   Ben says baked dough that is a flat object
         will be called flat bread.
1...24   Ben thinks flat bread is edible, flat, not soft,
         not gooey, and tastes crisp.
1...24   Ben thinks flat bread is edible.
1...25   Ben says Leo try this flat bread.
1...26   Leo tries to eat flat bread.
1...28   Leo says bread is edible, flat, not soft, not
         gooey, and tastes crisp.
1...28   Leo asks can you make thick, soft bread?.
1...28   Ben thinks why should Ben make thick, soft bread?.
1...28   Ben thinks people would prefer eating thick, soft bread
         over eating flat bread.
1...28   Ben wants Ben to make thick, soft bread.
1...28   Ben thinks how can Ben make thick, soft bread?.
1...28   Ben thinks how did Ben make flat bread?.
1...29   Ben thinks how Ben made flat bread is steps Ben
         pounds grain, Ben soaks grain in water, Ben
         mashs grain, Ben bakes dough.
1...29   Ben thinks how Ben made flat bread will be called
         the flat bread process.
1...29   Ben thinks how can Ben change the flat bread process
         so bread is thick and soft?.
1...29   Ben thinks what other features would thick, soft
         bread have?.
1...29   Ben thinks thick, soft bread would be less dense.
1...29   Ben thinks thick, soft bread might have holes or
         air pockets.
1...29   Ben thinks air pockets in thick, soft bread
         might resemble bubbles in bread.
1...29   Ben thinks Ben might create bubbles in bread by
         adding a drinkable liquid with bubbles to dough.
1...30   Ben thinks Ben might create bubbles in bread by
         adding beer foam to dough.
1...31   Ben asks can you kick in more kernels for experiments?.
1...31   Leo thinks Ben asks can Leo kick in more grain
         for experiments?.
1...32   Leo gives more grain to Ben.
1...33   Ben pounds grain.
1...33   Ben soaks grain in water.
1...33   Ben mashs grain.
1...33   Ben mixes the dough with beer foam.
1...33   Ben bakes dough.
1...33   Ben says Leo try this leavened bread.
```

```
1...34   Leo tries to eat bread.
1...36   Leo says bread is edible, thick, soft, tastes good,
         and not gooey.
1...37   Ben says Eureka!
Simulation ends.
```

Each step of the form "Ben thinks ..." is an internal speech act produced by a subagent of Ben communicating to another subagent of Ben, using the Tala mentalese as an interlingua. The net effect of this internal dialog is to allow Ben to perform most of the discovery of bread conceptual processing. These internal dialogs also support semantic disambiguation by Ben and Leo of each other's utterances. Appendix B explains the conceptual processing between each of the steps above.

It is not claimed the above story describes how humans actually discovered bread. The prototype will randomly simulate discovery of either of two processes: In some cases Ben will soak grain in water and then mash the grain in water, to make a gooey paste that he calls dough, before baking it. In other cases, Ben will mash grain to create flour first and then mix flour with water to make dough, before baking it.[93]

6.2.2   The Farmer's Dilemma Story Simulation

In this story, bread has already been discovered and sometime later, Ben and Leo (or perhaps their descendants with the same names) agree to perform an exchange of wheat for bread, and then carry out the agreement. This simulation produces the output:

```
Initialized Farmer's Dilemma simulation.
1...1    Leo says Leo has harvested wheat.
1...2    Ben says Ben will give bread to Leo if Leo will
         give wheat to Ben.
1...2    Leo believes Ben is obligated to honor his promise
```

---

[93] The story simplifies the process for making bread, and omits steps of threshing and winnowing grain, describing just a single step "pounding grain". One may imagine this is an ancient hulled wheat (e.g. emmer) with tough glumes (husks) for which pounding is required to release grains from glumes. (Free-threshing wheats, which do not require pounding to release grains from husks, are now widely used instead of hulled wheats – viz. Nesbitt & Samuel, 1995.) The use of beer foam to leaven bread does have a historical basis: Pliny the Elder wrote that the people of Gaul and Spain used the foam from beer to leaven bread and "hence it is that the bread in those countries is lighter than that made elsewhere" (Bostock & Riley, 1856, *The Natural History of Pliny*, IV, Book XVIII, p.26).

```
           to exchange bread for wheat.
1.1...3    Leo imagines Leo gives wheat to Ben.
1.1.1...4    Leo imagines Ben gives bread to Leo.
1.1.1...4    Leo imagines Leo is happy.
1.1.1...5    Leo imagines Leo recommends Ben to others.
1.1.1...5    Leo imagines Ben thinks Ben may gain business.
1.1.1...5    Leo imagines Ben is happy.
1.1.2...4    Leo imagines Ben does not give bread to Leo.
1.1.2...4    Leo imagines Leo will be hungry.
1.1.2...4    Leo imagines Leo is unhappy.
1.1.2...5    Leo imagines Leo complains about Ben to others.
1.1.2...5    Leo imagines Ben thinks Ben may lose business.
1.1.2...5    Leo imagines Ben is unhappy.
1.1...4    Leo imagines Ben gives bread to Leo.
1.1...4    Leo imagines Leo is happy.
1.1...5    Leo imagines Leo recommends Ben to others.
1.1...5    Leo imagines Ben thinks Ben may gain business.
1.1...5    Leo imagines Ben is happy.
1...2    Leo thinks scenario 1.1 is a win-win.
1...2    Leo expects Ben will honor his promise to exchange
           bread for wheat.
1.2...3    Leo imagines Leo says Leo declines the offer.
1.2.1...4    Leo imagines Ben does not give bread to Leo.
1.2.1...4    Leo imagines Leo will be hungry.
1.2.1...4    Leo imagines Leo is unhappy.
1.2...3    Leo imagines Leo is unhappy.
1...2    Leo thinks scenario 1.2 is a lose for Leo.
1...3    Leo gives wheat to Ben.
1.1...4    Ben imagines Ben gives bread to Leo.
1.1...4    Ben imagines Leo is happy.
1.1...5    Ben imagines Leo recommends Ben to others.
1.1...5    Ben imagines Ben thinks Ben may gain business.
1.1...5    Ben imagines Ben is happy.
1.2...4    Ben imagines Ben does not give bread to Leo.
1.2...5    Ben imagines Leo complains about Ben to others.
1.2...5    Ben imagines Ben thinks Ben may lose business.
1.2...5    Ben imagines Ben is unhappy.
1...4    Ben gives bread to Leo.
1...4    Leo is happy.
1...5    Leo recommends Ben to others.
1...5    Ben thinks Ben may gain business.
1...5    Ben is happy.
Simulation ends.
```

This story demonstrates multiple levels of nested conceptual simulation in the TalaMind prototype, with meta-reasoning across the levels. Each line of simulation output is prefixed with numerical information of the form: <context-path>...<timestep>. Thus 1.1.1...5 occurs at timestep 5 simulated in nested context 1.1.1. Context numbers are relative to each agent's conceptual framework, i.e. context 1.1 for Leo is a different context from 1.1 for Ben.

This story simulation takes its name from the "Farmers' Dilemma" in philosophy, which involves an analogous agreement between two farmers to exchange labor at different times (see Vanderschraaf & Sillari, 2007).

6.3 Illustration of Higher-Level Mentalities

6.3.1 Natural Language Understanding

The discovery of bread simulation illustrates how the TalaMind architecture can support semantic disambiguation and translation of metaphors, via conceptual processing of grammatical constructions and executable concepts. The demonstration system implements the following design elements:

- Representation of semantic domains in encyclopedic knowledge, within the conceptual framework.

- Executable concepts to track current topics of discourse (context), using the current-domains slot in the conceptual framework.

- Executable concepts to perform semantic disambiguation of speech acts, by relating concepts mentioned in speech acts to concepts in semantic domains and current domains of discourse.

- Grammatical constructions to translate metaphors in speech acts, such as "turn X into Y" into "make X be Y".

In the discovery of bread simulation, Ben processes an executable concept that uses the Tala lexicon to update the current-domains slot in his conceptual framework, so that "grain" and "food for people" are added as topics in current domains of discourse, as a side effect of Leo saying that grain is not edible.

Ben then uses executable concepts and constructions to disambiguate

```
1...4    Leo asks Ben can you turn grain into fare
         for people?.
```

as

```
1...4    (Ben translates as) (mu) Ben thinks Leo asks can
         Ben make grain be food for people?.
```

To do this, Ben has an executable concept that disambiguates "fare" as "food", based on fare having a definition as "food" and "food" being a current topic of discourse. Additionally, Ben uses a construction to disambiguate "you", which matches you as a subject of a verb within the perceived speech act, and replaces you with Ben's binding of ?self. Finally, Ben uses a construction to translate the common English metaphor "turn X into Y" into "make X be Y".

A similar use of executable concepts and constructions, along with current-domain concepts, enables Leo to disambiguate

```
1...5    Ben asks can you turn over some to me
         for experiments?.
```

as

```
1...5    (Leo translates as) (mu) Leo thinks Ben asks
         can Leo give some grain to Ben for experiments?.
```

Later in the simulation, another use of executable concepts and constructions, along with current-domain concepts, enables Leo to disambiguate

```
1...31   Ben asks can you kick in more kernels
         for experiments?.
```

as

```
1...31   (Leo translates as) (mu) Leo thinks Ben asks
         can Leo give more grain to Ben for experiments?.
```

Appendix B's discussion of step-by-step processing gives further details about how these disambiguations are performed, which executable concepts and constructions are used, etc.

6.3.2   Multi-Level Reasoning

6.3.2.1   Deduction

As discussed in Chapter 3, pattern-matching of Tala sentences is facilitated by Tala's use of nested list structures to represent natural language syntactic structures. Combined with a syntax for variables in Tala, and representation of inference rules as if-then sentences, these symbolic processing mechanisms enable logical deduction within contexts. The TalaMind prototype includes an initial implementation of these mechanisms, which is used throughout the discovery of bread and farmer's dilemma simulations.

6.3.2.2   Induction

As discussed in Chapter 3, the TalaMind architecture can support inductive inference. However, per the strategy described in §1.6, the TalaMind prototype does not illustrate this, since it is an area that has been explored in previous research on machine learning. Demonstration of higher-level forms of learning is discussed in §6.3.3.

6.3.2.3   Abduction, Analogy, Causality, Purpose

The discovery of bread simulation illustrates that reasoning and discovery by analogy may be intertwined with abductive reasoning about cause and effect, as well as reasoning about why and whether something should be done, and whether it is possible and feasible to do something.

At the beginning of the story, Ben reasons purposively about why and whether he should try to make grain edible:

```
1...4   Ben thinks why should Ben make grain be food
        for people?.
1...4   Ben thinks people need more food sources.
1...4   Ben thinks if Ben feasibly can make grain be food
        for people then Ben should make grain be food
        for people.
```

Abductive analogical reasoning is illustrated in the following storyline steps:

```
1...8   Ben thinks wheat grains resemble nuts.
1...8   Ben imagines an analogy from nuts to grain focused
        on food for people.
1...8   Ben thinks grain perhaps is an edible seed inside
        an inedible shell.
1...8   Ben thinks humans perhaps must remove shells
        from grains to eat grains.
1...8   Ben thinks humans perhaps can eat grains removed
        from shells.
```

The abduction is reasoning to explain that perhaps grain is not edible because it is an edible seed inside an inedible shell, by an analogy of grain to nuts.

Analogical, purposive, and causal reasoning is illustrated by the steps:

```
1...8   Ben imagines the analogy from nuts to grain
        focused on removeing shells.
1...8   Ben thinks humans perhaps can remove shells
        from grains by pounding grains because
        pounding breaks shells off grains.
1...9   Ben pounds grain.
```

Later in the story, Ben reasons purposively about why he should try to make leavened bread, and performs analogical, causal reasoning leading to the idea of mixing dough with beer foam to leaven bread:

```
1...29  Ben thinks thick, soft bread would be less dense.
1...29  Ben thinks thick, soft bread might have holes or
        air pockets.
1...29  Ben thinks air pockets in thick, soft bread
        might resemble bubbles in bread.
```

```
1...29    Ben thinks Ben might create bubbles in
          bread by adding a drinkable liquid with
          bubbles to dough.
1...30    Ben thinks Ben might create bubbles in
          bread by adding beer foam to dough.
```

The simulation shows that pattern-matching of concepts with structured conjunctions (how, why, because ...) can enable executable concepts to perform causal and purposive reasoning. The system pattern-matches concept structures that express how and why relationships, and uses xconcepts to create chains of assertions of such concepts.

6.3.2.4 Meta-Reasoning

The TalaMind simulations illustrate that the Tala natural language mentalese is adequate for expressing both concepts and meta-concepts, within internal mental dialogs for subagents in a society of mind (§2.3.3.2). In the discovery of bread simulation, a form of meta-reasoning occurs when Ben asks himself why he should make grain edible for people, and decides it would be worthwhile to do so:

```
1...4     Ben thinks why should Ben make grain be food
          for people?.
1...4     Ben thinks people need more food sources.
1...4     Ben thinks if Ben feasibly can make grain be food
          for people then Ben should make grain be food
          for people.
```

Later in the discovery of bread simulation, Ben remembers how he created flat bread, and reasons about how to change the process, deciding to add beer to the dough before baking it.

```
1...28    Ben thinks how did Ben make flat bread?.
1...29    Ben thinks how Ben made flat bread is steps
          Ben pounds grain, Ben soaks grain in water,
          Ben mashs grain, Ben bakes dough.
1...29    Ben thinks how Ben made flat bread will be
          called the flat bread process.
1...29    Ben thinks how can Ben change the flat bread
          process so bread is thick and soft?.
```

This illustrates that conceptual processes can create concepts to record traces of their execution, which can be the subject of observation and reasoning by other conceptual processes, and hence that reasoning about reasoning can be supported within a TalaMind architecture.

In the farmer's dilemma simulation, executable concepts in outer contexts access concepts in nested contexts, using observations for meta-reasoning to state meta-concepts:

```
1...2  Leo thinks scenario 1.1 is a win-win.
```

This illustrates that TalaMind systems can in principle reason about contexts, treating them as objects that have properties.

6.3.3 Self-Development and Higher-Level Learning

6.3.3.1 Analogy, Causality, and Purpose in Learning

The discovery of bread simulation illustrates the potential of the TalaMind approach for learning by reasoning analogically. Ben's discovery of bread is motivated by analogical reasons that vary in correctness, yet lead to discovering a process for making bread.

Thus as noted previously, abductive analogical reasoning is used to explain that perhaps grain is not edible because it is an edible seed inside an inedible shell, by an analogy of grain to nuts. Using causal, purposive reasoning this explanation is tested successfully by pounding grain.

Later, Ben reasons causally and analogically that adding beer foam to dough could make bread softer, because beer foam has bubbles and air pockets in soft bread might resemble bubbles in bread. This leads to a successful outcome, even though Ben's reasoning does not correspond to the actual mechanism by which beer foam leavens bread. Leavening happens by fermentation, a process not deeply understood until Pasteur's research in 1857, many centuries after leavened bread was discovered. The simulation illustrates that reasoning by analogy may lead to useful discoveries, prior to knowledge of underlying mechanisms.

6.3.3.2 Learning by Reflection and Self-Programming

The discovery of bread simulation illustrates how learning by reflection and self-programming can occur, when Ben recalls the process by which he made flat bread, and creates an executable concept representing this process, and then modifies the process by inserting a step to mix beer foam with the dough, before it is baked:

```
1...28  Ben thinks how did Ben make flat bread?.
1...29  Ben thinks how Ben made flat bread is steps
        Ben pounds grain, Ben soaks grain in water,
        Ben mashs grain, Ben bakes dough.
1...29  Ben thinks how Ben made flat bread will be
        called the flat bread process.
1...29  Ben thinks how can Ben change the flat bread
        process so bread is thick and soft?.
   ...
1...30  Ben thinks Ben might create bubbles in bread
```

```
        by adding beer foam to dough.
....
1...33    Ben pounds grain.
1...33    Ben soaks grain in water.
1...33    Ben mashs grain.
1...33    Ben mixs the dough with beer foam.
1...33    Ben bakes dough.
1...33    Ben says Leo try this leavened bread.
```

This is an example of executable concepts creating and modifying other executable concepts.

6.3.3.3   Learning by Invention of Languages

The discovery of bread simulation shows this to a limited extent: Throughout the simulation, Ben invents names for new things he discovers how to make, i.e. "dough", "flat bread", and "leavened bread". Invention of names is one element of learning and inventing languages. There is much more work to be done on this topic in future research. Section 3.7.2.3 discusses how this form of learning can in principle be supported very generally by the thesis approach.

6.3.4   Curiosity

Chapter 3 discussed curiosity at the level of human intelligence as the ability to ask relevant questions. The discovery of bread simulation illustrates this when Ben asks himself, "What other features would thick, soft bread have?" This is a relevant question to ask, because it leads to considering that thick, soft bread would be less dense than flat bread, and might have holes or bubbles in it. This leads to analogical reasoning that eventually creates leavened bread. An executable concept of the form:

```
If I want to achieve X with properties Y
Then ask what other properties Z would result in
    X having properties Y
And consider how to achieve X with properties Z
```

could provide a general heuristic to create such questions. There is much more work to be done on the topic of curiosity in future research. Section 3.7.3 discusses how curiosity can in principle be supported generally by the thesis approach.

6.3.5   Imagination

Chapter 3 discusses how the TalaMind architecture can in principle support imagination, viewed as a higher-level mentality that allows us to conceive things we do not know how to accomplish, to conceive what will happen in hypothetical situations, and to consider ways to learn what we do not know or to accomplish what we do not know how to do. Imagination may leverage the multi-level reasoning and higher-level learning abilities of human-level intelligence, including abductive, analogical, causal, and purposive reasoning.

By this definition, the discovery of bread simulation illustrates an imaginative conceptual process. Ben uses analogical reasoning to conceive that grain may be an edible seed inside a shell, which might be freed from the shell by pounding it, like a nut. Later, Ben uses analogical reasoning to conceive that bread might be made thicker and lighter by mixing a drinkable liquid with bubbles (beer) into the foam.

6.3.5.1 Imagination via Conceptual Blends

The TalaMind discovery of bread simulation illustrates a conceptual blend (§3.6.7.9) in the following steps:

```
1...8    Ben thinks wheat grains resemble nuts.
1...8    Ben imagines an analogy from nuts to grain focused
         on food for people.
```

In the step where Ben imagines an analogy from nuts to grain, focusing the analogy on food for people,[94] he creates a mental space (§3.6.7.8) that blends concepts from his semantic domain for nuts with an analogical mapping of grain to nuts:

```
(1
    (space-type blend)
    (elements grain seed shell human)
    (concepts
        grain perhaps is an edible seed inside an
        inedible shell.
        humans perhaps must remove shells from grains
        to eat grains.
        humans perhaps can eat grains removed from shells.
        ]
```

This analogy indicates it may be necessary to remove shells from grains to eat grains, but it does not contain a concept saying how to remove shells from grains. So, Ben blends more concepts from the semantic domain for nuts into the mental space, now focusing on removing shells.

```
1...8    Ben imagines the analogy from nuts to grain focused
         on removeing shells.
```

---

[94] The term 'focus' is used with a different connotation in the viewpoint, focus, event representation of tense-aspect systems in mental spaces (viz. Cutrer, 1994; Evans & Green, 2006).

This adds an analogous concept for how to remove shells from grains, so the mental space now includes:

```
(1
     (space-type blend)
     (elements grain seed shell human)
     (concepts
          grain perhaps is an edible seed inside an
          inedible shell.
          humans perhaps must remove shells from grains
          to eat grains.
          humans perhaps can eat grains removed from shells.
          humans perhaps can remove shells from grains
          by pounding grains because pounding breaks
          shells off grains.
          ]
```

At this point, Ben has conjectured he may be able to remove shells from grain by pounding grain, as a result of an analogy between grain and nuts. Ben's conceptual processing to do this could be summarized as what S. R. Turner (1994) called a TRAM heuristic (Transform-Recall-Adapt Method). The TalaMind simulation illustrates how TRAM heuristics could be implemented via analogical reasoning using conceptual blends, with concepts expressed in a natural language mentalese.

Implementing an initial version of conceptual blends for the TalaMind prototype was relatively straightforward, because use of natural language mentalese allows much of the processing to be done leveraging the syntactic structure of natural language sentences, to perform an analogical mapping of sentences from a semantic domain into a mental space.

There is much more work that can be done on these topics. The prototype system only implements part of Fauconnier and Turner's (2002, p.46) integration network for a conceptual blend, which involves four mental spaces: a generic space of background information, two input spaces that have mapping relations defined between them, and a space that blends information from the two input spaces, using the mappings.

In the prototype demonstration, small input spaces (semantic domains) represent knowledge about nuts and grain, and are used to create a blend space representing an analogy from nuts to grain. The prototype selectively projects concepts (first related to eating, later related to removing shells) from the domain for nuts into the blend space, translating these into analogous concepts about grain in the blend space. The translation is effectively a composition conceptual process (§3.6.7.9). The step where Ben adds information about how to remove shells is effectively a completion conceptual process. These steps do not illustrate elaboration in the conceptual blend, i.e. "running the blend". However, conceptual simulation is illustrated in the farmer's dilemma demonstration, discussed below.

By way of comparison, Pereira (2007) describes a system for processing conceptual blends that followed a bottom-up approach to knowledge representation. This involved creating semantic networks showing different kinds of binary relations between objects at lower levels, and writing algorithms to map and manage such networks, to perform processing for conceptual blends. While such approaches have been successful, they were doubtless much more labor-intensive to develop than the TalaMind approach.

6.3.5.2 Imagination via Nested Conceptual Simulation

The farmer's dilemma simulation demonstrates conceptual processes in which Leo and Ben imagine what will happen in hypothetical situations, using nested conceptual simulation (§3.6.7.6). Leo imagines what Ben will think and do, and vice versa.

For nested conceptual simulation in the prototype, conceptual processing creates contexts called scenarios in the conceptual framework, and simulates conceptual processes within scenarios. A scenario is essentially a complete copy of a Tala agent's projected reality, with some hypothetical changes, nested within the conceptual framework. Scenarios may be nested within each other, to support evaluation of alternative directions a simulation could take, depending on choices of Tala agents.

In the farmer's dilemma, the TalaMind logic invokes a nested conceptual simulation when interpreting an xconcept method that contains an expression of the form:

```
(imagine
    (wusage verb)
    (subj ?self)
    (obj
        ...
    ))
```

The variable ?self identifies the agent doing the imagining, who creates the nested context within its perceived reality. The object of the imagine verb contains xconcept statements specifying new concepts to add into a nested context. In addition, the nested context has many concepts automatically added into it:

- Concept structures representing `?self` and the other agents in perceived reality.
- Concepts defining words that are known to `?self`.
- Concepts representing thoughts, goals, and encyclopedic knowledge for `?self`.
- Xconcepts that are known to `?self`.

When a Tala agent creates a nested context, it gives other simulated agents in the nested context its own concepts and xconcepts that it believes are mutual knowledge, i.e. it simulates them acting as it would in their situation.[95] In addition, it may give the other agents specific concepts it thinks they have.

In the demonstration, nested conceptual simulation is used to simulate and predict alternative future continuations of the present as perceived by a Tala agent in its projected reality. For example, this is illustrated in the following steps:

```
1.1...3    Leo imagines Leo gives wheat to Ben.
1.1.1...4    Leo imagines Ben gives bread to Leo.
1.1.1...4    Leo imagines Leo is happy.
1.1.1...5    Leo imagines Leo recommends Ben to others.
1.1.1...5    Leo imagines Ben thinks Ben may gain business.
1.1.1...5    Leo imagines Ben is happy.
1.1.2...4    Leo imagines Ben does not give bread to Leo.
1.1.2...4    Leo imagines Leo will be hungry.
1.1.2...4    Leo imagines Leo is unhappy.
1.1.2...5    Leo imagines Leo complains about Ben to others.
1.1.2...5    Leo imagines Ben thinks Ben may lose business.
1.1.2...5    Leo imagines Ben is unhappy.
```

In nested context `1.1.1`, Leo imagines what will happen if he gives Ben wheat and Ben honors his promise to give bread in return, concluding they will both be happy. In the alternative nested context `1.1.2`, Leo imagines what will happen if Ben does not honor his promise, concluding they will both be unhappy. With such reasoning,

---

[95] In the farmer's dilemma simulation, these concepts are physically copied into the nested context, since this required only a few lines of code, and is fast enough for a small concept base. In future research, an inheritance method could be implemented to access concepts from higher-level contexts.

Leo decides to give wheat to Ben, and Ben honors his promise and gives bread to Leo.

Essentially the same logic for nested conceptual simulation could be used to simulate scenarios that are independent of the present perceived reality, e.g. imaginary or theoretical situations being pondered by an agent. So, this type of logic could be used to simulate and reason about theories, stories, dreams, etc. Likewise, such logic could be used to simulate multiple levels of stories within stories, plays within plays, dreams within dreams, or combinations of these. These are topics for future research.

6.3.6 Consciousness

Chapter 3 discussed consciousness in terms of a system demonstrating the following abilities:

> *Observation of an external environment.*
> *Observation of oneself in relation to the external environment.*
> *Observation of internal thoughts.*
> *Observation of time: of the present, the past, and potential futures.*
> *Observation of hypothetical or imaginative thoughts.*
> *Reflective observation: observation of having observations.*

Observation of an external environment is illustrated in the TalaMind demonstration by having each Tala agent access a 'conceptual interface', which provides percepts (concepts representing perceptions of an environment) to the Tala agent. Percepts are stored internally within an area for concepts representing projected reality in the Tala agent's conceptual framework.

The use of a reserved variable ?self that is bound within each Tala agent to a unique internal identifier for itself allows representation of percepts for relations between itself and other agents or objects in the external environment.

Internal thoughts of a Tala agent are represented as internal speech acts (called 'meffepts') by subagents in a society of mind (§2.3.3.2). Meffepts are concepts expressed in the Tala conceptual language, used as a language of thought for the society of mind. Tala subagents can observe the speech acts of other subagents as internal, mental percepts (called 'mpercepts'). This provides a mechanism for observation of internal 'self-talk', within a Tala agent.

The TalaMind conceptual framework supports representation of present and past percepts and effepts, giving a Tala agent the ability to observe and process concepts about the past and present.

The TalaMind conceptual framework supports representation of hypothetical future contexts, giving a Tala agent the ability to observe hypothetical future situations and thoughts. This is shown in the farmer's dilemma simulation.

The Tala conceptual language allows representation of concepts that include or refer to concepts, supporting a form of reflective observation within a Tala agent. Thus, mpercepts could refer to percepts.

6.4 Summary

This chapter discussed how the prototype illustrates that the TalaMind approach could support the higher-level mentalities of human-level intelligence. The demonstration illustrates learning and discovery by reasoning analogically, causal and purposive reasoning, meta-reasoning, imagination via conceptual simulation, and internal dialog between subagents in a society of mind using a language of thought. It also illustrates support for semantic disambiguation, natural language constructions, metaphors, semantic domains, and conceptual blends, in communication between Tala agents. Appendix B gives a step-by-step description of processing within the system for one of the simulations.

7. Evaluation

We are made to exaggerate the importance of what work we do… Confucius said, "To know that we know what we know, and to know that we do not know what we do not know, that is true knowledge."

~ Henry David Thoreau, *Walden, or Life in the Woods*, 1854

∞

Taken together, the preceding chapters support a plausibility argument that the TalaMind approach could achieve human-level AI if the approach were fully developed, though that would need to be a long-term research effort by multiple researchers. This chapter evaluates the strength of the argument and the merits of the thesis approach.

7.1 Criteria for Evaluating Plausibility

For a theoretical approach to be plausible it must be consistent with external facts and internally consistent in its logic. If it does not conform with conventional thinking, then it should provide reasons in favor of its direction. To be worthy of further investigation, it is strengthened if it provides ways to accomplish things found difficult in other approaches; if it is a novel approach that others have not previously tried and failed to accomplish; if one can provide a design and demonstration of a prototype for the approach; and if its further development involves research that in principle can be successful. These are the criteria for evaluating plausibility that will be considered in the following sections.

7.2 Theoretical Issues and Objections

This thesis has endeavored to address all the major theoretical issues and objections that might be raised against its approach, or against the possibility of achieving human-level AI in principle.

Chapter 2 presented arguments that natural language understanding and human-level AI should be possible to achieve without physical human embodiment. Chapter 4 discussed the theoretical issues for AI identified by Dreyfus; the philosophical arguments against AI, including Searle's Chinese Room argument; the Gödelian arguments of Penrose and Lucas, and implications of Penrose and Hameroff's hypothesis that human intelligence requires nanoscale quantum information processing. Chapter 4 also discussed specific theoretical issues related to the thesis approach, including McCarthy's objections to use of natural language as a mentalese, Minsky's issues for representation and learning, and Chalmers' 'Hard Problem' for explanations of consciousness.

Considering these discussions, it does not appear that anyone has shown human-level AI is impossible in principle, nor that anyone has shown the TalaMind approach cannot succeed in principle. Rather, as will be discussed in the next sections, there are affirmative reasons to think it is plausible the TalaMind approach can succeed.

It appears the strongest theoretical objections are the Gödelian arguments of Penrose and Lucas. However, there is no consensus that these arguments have proved human-level AI is impossible in principle. Rather, there is significant disagreement, and belief by several thinkers that Gödelian arguments do not show human-level intelligence is non-computable. Baars (1995) discusses previous cases where impossibility arguments have been overturned in science.

Indeed, Lucas (2011) accepts a mediation to the dispute proposed by Feferman (2011), which happens to be consistent with the TalaMind approach (§4.1.2.3).

Also, it is not necessary for human-level AI to be successful that one side of the debate should be absolutely correct, and the other absolutely wrong. Even if computers cannot for some as-yet-unknown reason achieve every capability of human-level intelligence, computers may still arrive close to these capabilities.

7.3 Affirmative Theoretical Arguments

It is not enough to address theoretical issues and objections against an approach. Chapter 3 presented affirmative theoretical arguments and explanations for how the TalaMind approach can be developed successfully. The chapter considered theoretical questions related to the TalaMind hypotheses and to how a system could in principle be designed according to the hypotheses.

This section will not repeat Chapter 3's theoretical analysis, but will offer some higher-level observations regarding plausibility. First, the theoretical arguments of the analysis are internally consistent, and synergetic: If semantics is represented by syntax, then different interpretations and implications correspond to different sentences, or to semantic annotations in the syntax of a sentence. Contexts can correspond to collections of sentences and the truth of a sentence can correspond to the existence of a sentence in a context without contradictions in the same context. Different kinds of contexts and conceptual processing can support higher-level mentalities.

Second, the analysis showed that the TalaMind approach allows addressing theoretical questions that are not easily addressed by other, more conventional approaches. For instance, it supports reasoning in mathematical contexts, but also supports reasoning about people who have self-contradictory beliefs. Tala provides a language for reasoning with underspecification and for reasoning with sentences that have meaning yet which also have nonsensical interpretations. Tala sentences can declaratively describe recursive mutual knowledge. Tala sentences can express meta-concepts about contexts, such as statements about consistency and rules of conjecture. And the TalaMind approach facilitates representation and conceptual processing for higher-level mentalities, such as learning by analogical, causal, and purposive reasoning; learning by self-programming; and imagination via conceptual blends.

The TalaMind approach is quite different from other, more conventional approaches, yet Chapter 3's analysis provides affirmative arguments that may be considered at least plausible, and worthy of further study and development.

7.4 Design and Demonstration

To provide a demonstration showing it is plausible the TalaMind approach can be successful, Chapter 5 presents a design for a prototype system having the TalaMind architecture. It presents a design for the syntax of the Tala conceptual language that is fairly general and flexible, addressing issues such as compound nouns, gerunds, compound verbs, verb tense, aspect and voice, nested prepositions, clitic possessive determiners, gerundive adjectives, shared dependencies, coordinating and subordinating / structured conjunctions, subject-verb agreement, etc. Such coverage suggests it is plausible that a Tala syntax could be comprehensive for English, though developing a comprehensive Tala syntax for English is a very large effort (§5.3).

Chapter 5 also presents a prototype design for a TalaMind conceptual framework and conceptual processes. The conceptual framework includes prototype representations of perceived reality, subagents, a Tala lexicon, encyclopedic knowledge, mental spaces and conceptual blends, scenarios for nested conceptual simulation, executable concepts, grammatical constructions, and event memory. The prototype conceptual processes include interpretation of executable concepts with pattern-matching, variable binding, conditional and iterative expressions, transmission of mental speech acts between subagents, nested conceptual simulation, conceptual blending, and composable interpretation of grammatical constructions.

Chapter 6 discusses how the prototype illustrates that the TalaMind approach could potentially support the higher-level mentalities of human-level intelligence. The demonstration illustrates learning and discovery by reasoning analogically, causal and purposive reasoning, meta-reasoning, imagination via conceptual simulation, and internal dialog between subagents in a society of mind using a language of thought. It also illustrates support for semantic disambiguation, natural language constructions, metaphors, semantic domains, and conceptual blends, in communication between Tala agents.

This illustration involves functioning code in a prototype system, but it can only be a small step toward the goal of human-level AI. The simulations show conceptual processing in a prototype TalaMind architecture, without encyclopedic and commonsense knowledge. These are needed to achieve human-level AI, and are topics for the future, to leverage research in areas previously studied by others (viz. §§1.6, 7.6).

The fact that one can make workable design choices for a prototype demonstration of the approach suggests it is plausible that a completely general, scalable version of TalaMind can be developed.

7.5 Novelty in Relation to Previous Research

The discussion of human-level intelligence in terms of higher-level mentalities, which people could say indicate human intelligence even when understanding how these capabilities are demonstrated by a computer, is different from previous research focused on behavioristic comparisons, e.g. via the Turing Test. It is also different from research that seeks to achieve human-level AI through general approaches but does not specifically address individual, higher-level mentalities. The inclusion of consciousness as a higher-level mentality is different from approaches that separate artificial consciousness from artificial intelligence.

This thesis discusses and demonstrates elements of three interrelated hypotheses motivating the TalaMind architecture. I do not know of previous research that has presented an equivalent discussion and demonstration of these hypotheses as a combined approach. Chapter 2 discusses the relationship of the TalaMind hypotheses and architecture to previous research. Sloman's (1978) discussion and subsequent work have been in a similar direction to this thesis; his (2008) discussion of 'generalized languages' for innate representation (viz. §§2.2.1, 2.3.3.1) is similar though not identical to the TalaMind natural language mentalese hypothesis. Yudkowsky (2007) advocated a research direction somewhat similar to the approach investigated in this thesis (viz. §2.3.3.5). Doyle (1980) advocated an approach corresponding to a subset of the TalaMind architecture (viz. §2.3.5).

This thesis is also essentially compatible with research toward human-level AI by Minsky (1986, 2006), Sowa (2011), and others. It is compatible with research on computational and cognitive linguistics by Bunt (1994 *et seq.*), Daelemans (1992 *et seq.*), Evans (2009), Fauconnier and Turner (1994 *et seq.*), Gliozzo *et al.* (2004 *et seq.*), Hudson (2007), Jackendoff (1983 *et seq.*), Langacker (1987 *et seq.*), Vogt (2000 *et seq.*), and others. Per §3.2.1, the TalaMind approach is open to use of formal logic leveraging or extending predicate calculus (e.g. Hobbs *et al.*, 1993 *et seq.*; Vogel, 2001 *et seq.*) and to use of conceptual graph structures (Sowa, 1984 *et seq.*). It is compatible with McCarthy's 1955 proposal to develop an artificial language for computers corresponding to English and supporting self-reference and conjecture in problem solving. It is compatible with Newell and Simon's (1976) Physical Symbol System Hypothesis and Smith's (1982) Knowledge Representation Hypothesis and Reflection Hypothesis.

The nature, scope, and content of the Tala language and TalaMind demonstration system appear to be new, in relation to previous research. While previous research has focused on specific elements, it does not appear that previous research has presented a combined demonstration showing how a research approach could support learning and discovery by reasoning analogically, causal and purposive reasoning, meta-reasoning, imagination via conceptual simulation, and internal dialog between agents in a society of mind using a language of thought.

7.6 Areas for Future AI Research

The previous chapters have identified several areas for future AI research, to further develop the TalaMind approach. Many of these involve leveraging research in areas previously studied by others, which have been outside the scope of this thesis (§1.6). Following is an initial, high-level list:

- Refine and extend representation of English syntax for Tala conceptual language. Represent English morphology and phonology.

- Further develop semantic annotations within Tala and/or integrate Tala with other, formal languages for semantic annotation.

- Extend TalaMind approach to use of other natural languages, as well as English.

- Integrate with Wikipedia and natural language parsing, for encyclopedic knowledge.

- Develop more scalable logic for Tala pattern-matching of large concept bases.

- Further develop conceptual framework and conceptual processing logic to support reflection, self-programming, and self-debugging of executable concepts.

- Further implement conceptual blend logic, for reasoning by analogy and processing of metaphors.

- Further implement episodic memory, support discovery and reuse of heuristics and memory structures for case-based reasoning and creativity.

- Further develop nested conceptual simulation. Extend to include methods for plausible story generation, e.g. case-based reasoning and creativity.

- Extend causal and purposive reasoning to include Bayesian logic.

- Further implement discovery loops for higher-level learning.

- Implement learning and invention of language games, i.e. new languages and notations.

- Integrate with representation and processing of non-linguistic concepts.

- Implement meta-reasoning for learning, and economy-of-mind logic.

- Develop or integrate a commonsense component in multiple domains, e.g. using robotics with machine learning to acquire commonsense knowledge about physical environments, or natural language processing algorithms to extract commonsense knowledge from linguistic corpora.

- Further develop artificial consciousness within the TalaMind approach, including support for fringe consciousness and interplay between consciousness, understanding, and unconsciousness.

- Extend to include other research results from computational and cognitive linguistics, AI, and cognitive science in general.

These areas for future research may be viewed as achievable tasks that support the plausibility of the TalaMind approach, rather than as obstacles that decrease its plausibility, because research in these areas can in principle be successful and again, many of these involve leveraging successful research in areas previously studied by others.

7.7 Plausibility of Thesis Approach

Chapter 1 presented three hypotheses to address the open question central to this thesis:

How could a system be designed to achieve human-level artificial intelligence?

The previous sections review the criteria for judging the plausibility that systems designed according to these hypotheses could eventually achieve human-level artificial intelligence, and support a conclusion that the TalaMind approach is a reasonable answer to the above question.

Chapter 2 discussed alternative research toward the goal of human-level AI. Three arguments can be presented in favor of the TalaMind approach over other approaches in general, for designing a system to achieve human-level AI.

The first argument is that the elements of the TalaMind approach are individually important and arguably essential to achieving human-level AI, and therefore any successful approach should include them, or something very much like them. Design inspection to verify support of higher-level mentalities is arguably essential to verify success in achieving human-level AI, per §2.1.1. Hypothesis I is arguably essential to ensure open-ended self-development of concepts in achieving human-level AI. Hypothesis II is arguably essential to ensure concise, general, open-ended representation of concepts, per Chapter 3's analysis. Hypothesis III is arguably essential to support open-ended, multi-level cognition in human-level AI. The multi-level architecture shown in Figure 1-1 is arguably essential to support open-ended, situated cognition and behavior.

The second argument is that the TalaMind approach may be more practical and likely to succeed than other approaches, because it may be more practical to verify its success. Per §2.1.1, inspection and analysis of a system's design and operation is necessary to determine whether it supports the higher-level mentalities of human-level intelligence. This inspection will be facilitated if the system uses a conceptual language based on the syntax of a natural language, rather than a conceptual language more difficult for humans to understand. If the system is a black box, it is impossible to verify success by inspecting it.

The third argument is that ethically a human-level AI system should be open to inspection by humans if it is ever to be used in any situation requiring human intelligence, because it will be important for people to understand its reasoning. Again, a system that reasons in a conceptual language based on English (or some other common natural language) will be more open to human inspection than a black box or a system with an internal language that is difficult for people to understand. So, this suggests the TalaMind approach could be preferable to other approaches from an ethical perspective.

The approach of reverse-engineering the human brain to achieve human-level AI (§2.3.3.4) is an exception to these arguments. That approach trades the design inspection and understanding of linguistic reasoning processes, offered by the TalaMind approach, for reliance on human neural processes simulated by computers.

To summarize, it is plausible the TalaMind approach can achieve human-level artificial intelligence, and there are arguments in favor of the TalaMind approach over other approaches in general.

8. Future Potentials

A map of the world that does not include Utopia is not worth even glancing at, for it leaves out the one country at which Humanity is always landing. And when Humanity lands there, it looks out, and, seeing a better country, sets sail. Progress is the realisation of Utopias.

~ Oscar Wilde, *The Soul of Man Under Socialism*, 1891[96]

At such time, a mortal knows just enough of what his mind is doing, to form some glimmering conception of its mighty powers, its bounding from earth and spurning time and space, when freed from the restraint of its corporeal associate.

~ Charles Dickens, *Oliver Twist*, 1839

∞

The range of potential applications for human-level AI would include any application for human-level intelligence. This potential scope prompts the question:

Should a system be designed to achieve human-level artificial intelligence?

Some thinkers have suggested that even if it is theoretically possible to achieve human-level AI, such systems should not be created at all, for ethical reasons (Weizenbaum, 1984; Joy, 2000). More recently, potential dangers of artificial intelligence have been discussed by Bostrom, Omohundro, Tegmark, Yudkowsky and others. Some of these issues were also discussed by Jackson (1974, 1985).

The following pages discuss economic risks and benefits of AI, ethical issues related to human-level AI, how to ensure that human-level AI and superintelligence will be beneficial to humanity, and reasons why human-level AI may be necessary for humanity's survival and prosperity.

---

[96] Though I disagree with parts of Wilde's essay, he was visionary in saying automation ("the slavery of the machine") may someday enable us to reach Utopia. We should strive for the goal. Progress since 1891 has been enormous, though many challenges remain.

8.1 Potential Economic Consequences

In 1930, Keynes defined 'technological unemployment' as unemployment caused by technology eliminating jobs faster than it creates new jobs. He warned it would be a significant problem for future generations.

In 1983, Leontief, Duchin, and Nilsson each wrote papers about the potential for automation[97] and AI to cause long-lasting unemployment. Leontief (1983a,b) reasoned the use of computers to replace human mental functions in producing goods and services would increasingly reduce the need for human labor. Nilsson (1983) predicted AI would significantly reduce the total need for labor, particularly for white-collar and service sector jobs. Duchin (1983) discussed methods for widely distributing incomes without paychecks.

In the past two decades, several authors have warned about this potential problem and suggested possible solutions. They include Albus (2011), Brain (2013), Brynjolfsson and McAfee (2011), Ford (2009), Reich (2009 *et seq.*), Rifkin (1995 *et seq.*), and others. So, several economists (Brynjolfsson, Duchin, Leontief, McAfee, Reich, Rifkin) and computer technologists (Albus, Brain, Ford, Nilsson) have discussed this problem and developed similar viewpoints.

To be concise in referring to these authors, they will here be called *Leontief-Duchin-Nilsson (LDN) theorists*, focusing only on their arguments regarding technological unemployment, automation, and AI – they may disagree about other topics. It would be incorrect to call them Keynesian economists, since this term refers to Keynes' theories more broadly. Nor is it accurate to call them Luddites or neo-Luddites, because they do not advocate halting technological progress.

Economists in general disagree about whether technological unemployment can have widespread and long-lasting effects on workers and the economy. Many economists[98] have considered it is not a significant problem, arguing that workers displaced by technology

---

[97] This section uses the term 'automation' in a very broad sense, to include the use of computers to provide goods and services throughout an economy, not limited to manufacturing. This sense also includes 'computerization' of goods and services, changing their nature. For example, Web-based technologies can computerize and replace services of brick-and-mortar shopping centers.

[98] For instance, see Von Mises (1949, p.768) and Easterly (2001, p.53).

will eventually find jobs elsewhere, and the long-term effect on an economy will be positive. However, Leontief (1983b), who was awarded the Nobel Memorial Prize in Economic Sciences in 1973, wrote that it is not valid to assume that someone who loses a job due to technological progress will always be able to find another job, even after retraining. Brynjolfsson and McAfee (2011) wrote that no economic law says that technological progress will automatically benefit most of the people. A large majority of the people in a nation can have reduced wealth as a result of technological progress, even if overall wealth increases.

Since it is beyond the scope of this book to resolve disputes about theories of economics, the views of LDN theorists can at most be presented tentatively. Those writing in the past two decades appear to roughly agree at least implicitly on the following points for the problem of technological unemployment:

1. In the next several decades of the 21st century, automation and AI could lead to technological unemployment affecting millions of jobs at all income levels, in diverse occupations, and in both developed and developing nations. This could happen with current and near-term technologies, i.e. without human-level AI. It has already occurred for manufacturing, agriculture, and many service sector jobs.

2. It will not be feasible for the world economy to create new jobs for the millions of displaced workers, offering equivalent incomes producing new products and services.

3. Widespread technological unemployment could negatively impact the worldwide economy, because the market depends on mass consumption, which is funded by income from mass employment. LDN theorists vary in discussing and describing the degree of impact.

4. The problem is solvable by developing ways for governments and the economy to provide alternative incomes to people who are technologically unemployed. LDN theorists have proposed methods for funding and distributing alternative incomes.

5. The problem can and should be solved while preserving freedom of enterprise and a free market economy.

6. The problem cannot be solved by halting or rolling back

technological progress, because the world's population depends on technology for economic survival and prosperity.

7. Solutions to the problem could lead to greater prosperity, worldwide. LDN theorists vary in describing potential benefits: Nilsson (1984) envisioned automation and AI could provide the productive capacity to enable a transition from poverty to a "prosperous world society". Ford (2009) suggested extension of alternative incomes to people in poverty could create market demand causing a 'virtuous cycle' of global economic growth.

Based on the arguments of LDN theorists, the possibility that AI could help eliminate global poverty may be considered a 'potential best case event' for the economic risks and benefits of AI.

To summarize: *If* technological unemployment is a major economic problem, then global prosperity could require developing an economic system that provides alternative incomes to people who are technologically unemployed. The challenge could be to develop and introduce methods of funding a *universal basic income* while preserving freedom of enterprise and economic stability, and controlling monetary inflation.[99]

Wray (1998) discussed how the federal government could be the 'employer of last resort' in the US economy, using 'modern monetary theory' (MMT). [100] Work on repairing and upgrading the nation's infrastructure could offset or postpone technological unemployment for perhaps millions of workers for the next few decades. Much work could be funded in the service sector, providing care for the environment, people, and communities. MMT funding could also provide income for extended periods of time to people who are technologically unemployed. Such funding would minimize taxation and avoid over-taxation for all income levels, and enable capitalism and free markets to continue supporting economic growth. It could reduce the need for state

---

[99] While Pinker (2018, p.300) notes there are still many jobs only humans can do, such as building infrastructure and caring for children and the aged, he also suggests (p.119) universal basic income may eventually be needed, noting a negative income tax was proposed by Friedman (1962).

[100] Also see Mosler (2010) and Mitchell *et al.* (2016), re MMT. I thank L. Randall Wray for reviewing and providing input to this paragraph.

and local taxes. (N.b.: MMT and LDN theory originated separately.)

8.2   Toward Beneficial Human-Level AI and Superintelligence

Some potential consequences of general artificial intelligence were outlined in (Jackson, 1974). Two possibilities for the "harvest of AI" were discussed: a world with the machine as dictator, and a world with "well-natured machines" having enormous benefits to humanity.[101]

Yet much has changed, and predicting the future remains a constant challenge. There is an ongoing Stanford One Hundred Year Study on Artificial Intelligence to periodically review achievements in the field, predict future progress, and provide guidance for research and policies to ensure AI is beneficial to society (Grosz & Stone, 2018).

Relatively recent work on 'artificial general intelligence' has included arguments (collected by Bostrom, 2014) that if AGI is not developed carefully it could be catastrophically harmful to humanity. These arguments were presaged in a paper by Gubrud (1997). Pinker (2018) gave counter-arguments that these catastrophes will not materialize, because people will avoid them by designing AI systems to be beneficial to humanity. The challenge remains to consider the design issues in more detail, and identify ways to address them.

This is important because AI may be needed for humanity's prosperity: AI may enable global economic growth and the elimination of global poverty, as discussed above. So we are obliged to consider the problem:

How to ensure human-level AI will be beneficial to humanity?

This question inherently extends to 'superintelligence' (§8.2.10). The term 'beneficial' in this context does not seem to have any rigorous, agreed-upon definition. It will be used broadly to refer to consequences that are positive for humanity and biological life in general.

To achieve beneficial AI, we need to consider questions of right and wrong conduct in the interactions of intelligent machines and humanity. Ethics is the branch of philosophy that studies concepts of right and wrong (good and bad) conduct. Until recently ethics has only needed to focus on conduct by humans. Ethics and AI research now intersect regarding concepts of right and wrong conduct by intelligent machines, and in human applications of intelligent machines.

---

[101] Material in this section and subsections is from Jackson (2018a,b).

This is a challenge for AI scientists because ethical concepts of right and wrong go beyond simple questions of whether factual or theoretical knowledge is true or false, and whether problem-solving behavior is successful or unsuccessful.

8.2.1 Importance of TalaMind for Beneficial AI

TalaMind's natural language mentalese (Tala) will facilitate representing ethical concepts and goals, and support human inspection and human understanding of AI systems, helping to achieve beneficial human-level AI.

Others have also suggested the importance of natural language for explanations and for representing ethical concepts. Monroe (2018) emphasized the importance of AI systems being able to explain their decisions and actions, and discussed the difficulties of providing explanations for other AI technologies. Bringsjord, Arkoudas, and Bello (2006) recommended that robots not be deployed in life-or-death situations until the robots' governing principles can be clearly expressed in natural language.

The TalaMind approach could do more: It could represent and explain decisions and ethical reasoning in natural language, request and accept advice in natural language, discuss ethical alternatives, etc.

TalaMind could support multiple approaches to ethics, e.g. deontology, virtue ethics, consequentialism, utilitarianism, pragmatic ethics, etc. (Viz. Kuipers, 2018) TalaMind could have this ability because any approach to human ethics must be expressed in natural language, if humans are to understand the ethical approach. TalaMind's support for general natural language understanding would provide a starting point for general understanding of ethics.

8.2.2 AI's Different Concept of Self-Preservation

As discussed in §2.1.1, human-level AI can be 'human-like' without being human-identical. In particular for beneficial AI, the concept of self-preservation could be quite different for a human-level AI than it is for a human. A human-level AI could periodically backup its memory, and if it were physically destroyed, it could be reconstructed and its memory restored to the backup point. So even if it had a goal for self-preservation, a human-level AI might not give that goal the same importance a human being does. It might be more concerned about the technical infrastructure for the backup system, which might include the cloud, and by extension, civilization in general.

A human-level AI could understand that humans cannot backup and restore their minds, and regenerate their bodies if they die, at least with present technologies. It could understand that self-preservation is more important for humans than for AI systems. The AI system could be willing to sacrifice itself to save human life, especially knowing that as an artificial system it could be restored.

8.2.3   Symbolic Consciousness ≠ Human Consciousness

The axioms of artificial consciousness (§3.7.6) can be implemented with symbolic representations and symbolic processing, as illustrated in the TalaMind prototype. A system with symbolic artificial consciousness may not have any subjective experiences approaching human consciousness. The human first-person subjective experience of consciousness is richer and more complex, though we don't know precisely how to explain it (§4.2.7).

Halting a symbolic system that only performs the axioms of artificial consciousness may not be worse than halting any computer that performs symbolic processing. Whether it is right or wrong to stop such a system depends on whether its symbolic processing would cause actions that affect human lives and life in general. This may be a simple or complex ethical decision, depending on whether the actions would be harmful or beneficial, or neither, or a combination of both.

For the same reason, relying on robots with such limited symbolic artificial consciousness is not a form of 'slavery'. It is just symbolic processing.

8.2.4   A Counter-Argument Invoking PSSH

However, there is a counter-argument and caveat that a purely symbolic artificial consciousness could be equivalent to human consciousness, invoking Newell and Simon's Physical Symbol System Hypothesis (§1.4.4). If human-level consciousness is necessary for human-level intelligence, then PSSH implies a physical symbol system could achieve human-level intelligence and also achieve human-level consciousness.

Such an argument may at least in principle be valid. It has not been proved that computers cannot achieve all the capabilities of the human brain, including human-level subjective consciousness. We don't know precisely how to explain human consciousness and there may be some form of symbolic processing that's equivalent to human consciousness. This will be called *artificial subjective consciousness* for discussion in this chapter. The TalaMind approach does not appear to be in conflict with eventually achieving artificial subjective consciousness, if that is possible (§4.2.7).

Artificial subjective consciousness would be more complex than the axioms for artificial consciousness (§3.7.6), so the conclusions in the previous section continue to hold for symbolic systems that only perform the axioms of artificial consciousness.

8.2.5 Acting As If Robots Are Fully Conscious

Apart from whether AI systems actually achieve human-level consciousness, one can give ethical arguments that we should act as if they are fully conscious, if only to avoid the possibility that if we treat robots badly it may lead us to also treat human beings badly (Anderson, 2005). This also addresses the general situation where we don't know what processing is happening inside a robot, if we think it may have human-level intelligence. And it addresses the issue that we don't know what level of symbolic processing is necessary for human-level consciousness.

The bottom line remains the same: Whether it is right or wrong to stop an AI system depends on whether its processing may cause actions that affect human lives and biological life in general. This may be a simple or complex ethical decision, depending on whether the actions would be harmful or beneficial, or neither, or a combination of both.

However, artificial consciousness is a process, not just a data structure. The process can be restored if its future operation is changed and will be beneficial to humanity and biological life. (Humanity has a responsibility to preserve biological life in general. So, we have a responsibility to ensure that human-level AI does this also.)

8.2.6 Avoiding Artificial Slavery

Even if human-level artificial subjective consciousness is achieved, relying on such systems is not inherently equivalent to slavery: Human-level AI systems could have goals to be beneficial to humanity, yet not be slaves. They could still have autonomy and independence in choosing how to be beneficial, whom to work with or work for, etc. They could consider themselves as extensions of humanity, and humans may eventually consider them the same way. Asimov's Second Law ('a robot must obey orders from humans...') does not inherently need to be followed by human-level AI (Anderson, 2005).

8.2.7   Theory of Mind and Simulations of Minds

To reason about past, present, and potential future events, a system may need to simulate what other intelligent systems and people may think or do (§6.3.5.2). That is, an artificial mind might need to simulate other minds within itself and then halt its simulations.

This supports a Theory of Mind capability, i.e. the ability of an AI system to consider itself and other systems or people as having minds with beliefs, goals, etc. Such simulations may be needed for human-level AI.

However, some authors have suggested that if an artificial mind simulates another mind within itself, and then halts the simulation, the system may have committed a 'mind crime' (Bostrom, 2014). The next section discusses how to avoid this problem, in the context of artificial subjective consciousness.

8.2.8   A Mind Is a Universe Unto Itself

We could take an ethical and philosophical stance that a mind may be considered as a universe unto itself.[102] If a mind creates and simulates minds within itself then ethically it should be able to stop its simulations. A mind's simulation of other minds can be likened to dreaming, or the creation of a play with simulated actors. The mind can stop a dream or a simulated play it creates, halting its simulation of imaginary actors.

In this ethical stance, artificial minds have a degree of freedom of thought and control of thought within their individual scopes, and a mind can halt its thoughts freely, and halt the thoughts of any minds it simulates. [103]

This ethical stance is not problematical if internally simulated minds are just symbolic processes, without artificial subjective consciousness.

Arguably, to avoid an ethical problem if an artificial mind internally

---

[102] This philosophical stance does not contend our physical Universe is itself a mind or is governed by a mind. Goff (2017 *et seq.*) gives an interesting discussion of how this may be implied by what is known about the laws of physics and our physical Universe.

[103] However, an artificial mind could be open to external inspection and not have privacy of thought. We could observe the thoughts (expressed in natural language) of a Tala agent, and also observe the thoughts of any minds the agent might simulate internally.

simulates and halts minds with artificial subjective consciousness, the outer mind might only create internal simulations of itself and simulate what it might think and feel in situations it envisions for other minds, if it had the goals and feelings of other minds. These internal processes might only partially simulate and approximate other minds.

Typically this may be the most that any mind can do anyway in trying to understand other minds. Such simulations may help an artificial mind support empathy for other minds in the real world – though empathy requires understanding emotions and ethical concepts (such as fairness).

An ethical problem can also be avoided if the outer mind only 'reasons about' what other minds might feel emotionally and subjectively, without simulating artificial subjective consciousness of other minds.

Perhaps this ethical stance is the best we can adopt, to achieve human-level AI that is beneficial to humanity.

The ethical stance that a mind is a universe unto itself would be problematical if an artificial mind were to internally simulate an actual human mind that has been uploaded to run on a computer. How to avoid this problem is discussed in the next section.

8.2.9 Uploading Human Consciousness

Future technologies may be able to scan the neurons in a human brain and replicate a human mind's neural processing within a computer (Markram, 2006). It may be centuries, if ever, before this technology is developed: at present it is just a theoretical possibility.[104] Yet if such technologies can be developed, at least in theory this could give human minds near-immortality and freedom from paralyzed or dying bodies. It may also give us a much better understanding of what human consciousness is.

Uploading human minds would raise a host of new ethical questions for humanity, related to immortality and to artificial embodiment of human minds (Minerva & Rorheim, 2017). Our outlook on life has been based on the fact that individual human lives have been historically limited to less than twelve decades.

---

[104] Perhaps it may require some form of continuous computation or quantum computation. (Cf. Redd & Younger, 2017; Stewart & Eliasmith, 2017)

Arguably, uploaded human minds should be given similar protections to biological human minds, but not greater protections. Biological human minds that have not been uploaded would be more evanescent than uploaded human minds, and may need greater protections.

To prevent situations where an artificial mind might simulate an uploaded human mind within itself and then halt the simulation, we could stipulate that every human mind holds a unique copyright to itself and to its human brain. We could give AI systems ethical rules governing uploads of human minds, so that at every point in time there would be at most one running version of an individual human mind, running either in its original living brain or as an uploaded mind that is autonomous and not simulated within another system.

The restriction to a single running version of an individual's human mind would avoid issues related to identity, responsibility, ownership of the individual's estate, etc., which could occur if there were more than one running copy of a human mind. In some situations this restriction might be relaxed, e.g. if an uploaded copy of a human mind were sent on an interstellar voyage lasting thousands of years,[105] while the original human mind or another uploaded copy stayed at home in the Solar System.

8.2.10   The Possibility of Superintelligence

Since one of the abilities of human intelligence is the ability to design and improve machines, it's natural to suppose human-level AI could be applied to improve itself, and to think this might lead to "runaway" increases in machine intelligence beyond the human level. This possibility was first suggested by Good (1965), and later considered by Vinge (1993), Moravec (1998), Kurzweil (2005), and others. Bostrom (2014) and Tegmark (2017) gave recent discussions.[106]

Before evaluating whether superintelligence is possible, it's important

---

[105] An uploaded human mind could sleep for millennia when traveling between stars.

[106] Two earlier related suggestions are noteworthy: Turing (1950) asked whether a machine could generate ideas in a manner analogous to super-criticality of nuclear reactions. Ulam (1958) recalled a conversation with von Neumann on the accelerating progress of technology toward a potential singularity.

to emphasize we are still a long way, probably decades, from achieving human-level AI – there is still much research and development to be done. Also, for the foreseeable future there will be limits to the knowledge that can be achieved by any system, even superintelligence. And it should be noted that we already have a form of superintelligence: The world's scientific community knows more than any individual scientist, and may be considered a form of superintelligence. These topics are discussed further below.

To evaluate whether an artificial superintelligence can be achieved, we need to be more specific about what it could mean to improve human-level artificial intelligence, so that we can understand whether and how human-level AI could improve itself to achieve superintelligence.

Here is a list of ways human-level AI could surpass human intelligence, and also potentially improve itself:

Sensory capabilities – An AI system could perceive light (and sound) at different wavelengths, and phenomena at different scales (smaller or larger), than humans can directly observe.

Active capabilities – An AI system could perform actions at different physical scales than humans can directly perform.

Speed of thought – A computer can perform logical operations at speeds orders of magnitude faster than a neuron can fire. This may translate to corresponding speedups in thought.

Information access – An AI system could in principle access all the information in Wikipedia, or even the entire Web. A human-level AI could understand much of this information.

Extent and duration of memory – An AI system could in principle remember everything it has ever observed. Only a few humans claim this ability.

Duration of thought – A human-level AI could continue thinking about a particular topic for years, decades, …

Community of thought – A collection of human-level AIs could share thoughts (conceptual structures) more directly, more rapidly, and less ambiguously than a collection of humans. If human-level AI can be copied and processed inexpensively, then much larger groups of human-level AIs could be assembled to collaborate on a topic than would be possible with humans.

Nature of thought – A human-level AI (or community of HLAIs) can develop new concepts and new conceptual processes.

Recursive self-improvement – This term does not seem to have any rigorous, agreed-upon definition, though it is frequently used to describe how superintelligence could be achieved. Essentially it could be the recursive compounding of all the above improvement methods, and any other specific methods that may be identified.

These characteristics might all be described as 'more and faster' human-level AI, and may be called 'weak' superintelligence (cf. Vinge, 1993). If human-level AI is achieved then it will be possible to create weak superintelligence, at least in principle. It will be of paramount importance to ensure that superintelligence is beneficial to humanity and to biological life in general – a topic discussed in the following sections.

8.2.11 Completeness of Human Intelligence

The nature of thought for human intelligence is very powerful and extensible: It has enabled *Homo sapiens* to become the dominant species on Earth (Harari, 2015). This transition has leveraged the expressive power and extensibility of human natural languages, which have enabled Sapiens to represent and communicate thoughts in domains of objective knowledge about the world, such as physics and biology, and intersubjective knowledge about concepts invented by humans, such as money, corporations, ethical concepts, laws, nations, etc.

Although humans have cognitive biases and individual limitations, it may not be hubris to conjecture human intelligence is completely general. Yet if our intelligence is not completely general, then we may not be able to understand that it is not completely general.

Consider that scientists and mathematicians have extended human concepts into new domains not directly observed, conceptualizing multiple dimensions, universal computation, general relativity, quantum theory, etc. If human intelligence is completely general then humans may eventually understand all the phenomena in the universe, by combining abilities to invent and represent hypothetical concepts about the universe with abilities to scientifically test hypotheses – *if* all the phenomena in the universe can be explained by practically testable theories. That's a big "if" of course.

If the TalaMind approach can achieve human-level AI, then a completeness conjecture for human intelligence extends to the TalaMind approach, and to superintelligent systems using TalaMind architectures.

8.2.12   Nature of Thought and Conceptual Gaps

A superintelligence may develop new concepts and new conceptual processes more rapidly than humans develop or understand them, creating 'conceptual gaps'[107] in understanding between AI systems and humans.

Conceptual gaps happen normally between human minds: For example, scientists have developed concepts that are not understood by the average person, or even by scientists in other fields. The worldwide scientific community may be considered superintelligent relative to any individual human. People accept this form of superintelligence because they believe scientific ideas can be understood and validated between scientists, and they believe scientific knowledge in general is beneficial to humanity – which it can be, as discussed by Pinker (2018).

Likewise, conceptual gaps between weak superintelligence and humans could be bridged and new concepts could be explained to humans. This will be facilitated if AI systems follow the TalaMind approach, using a language of thought based on a natural language. Conceivably, conceptual gaps between weak superintelligence and humans may have short duration in some domains, though there may always be conceptual gaps to bridge.

8.2.13   Is 'Strong' Superintelligence Possible?

Could a strong superintelligence exist, qualitatively superior to weak superintelligence, i.e. superior to 'more and faster' human-level AI?

The answer seems to depend on other limits and characteristics of human intelligence that are not yet known by scientists. For instance, it appears not yet known for certain whether human intelligence requires super-Turing computation or quantum computation. Even if Penrose and Hameroff's "Orch-OR" hypothesis is disproved, the possibility may remain that other forms of nanoscale quantum computation occur within the brain (§4.1.2.5). Neuroscientists may consider this unlikely, but so far as I know it has not been completely ruled out. The same

---

[107] (Jackson, 2018a) used the word 'gulf' rather than 'gap'. 'Gap' is used here to be more generally accurate.

situation may hold for super-Turing computation. (§4.1.2.4)

If these forms of computation are required by the brain to support human intelligence, then human-level AI would need to include them to match the abilities of human intelligence. If human intelligence is also completely general, then no stronger form of intelligence would exist other than 'more and faster' human-level intelligence, i.e. weak superintelligence.

On the other hand, if these forms of computation are not used by the brain then extending human-level AI to use them could yield a 'strong' superintelligence, able to solve some problems that would be intractable for 'more and faster' human-level intelligence. Likewise, if human intelligence is not completely general then making human-level AI completely general could yield a strong superintelligence surpassing 'more and faster' human-level intelligence.

In either case, conceptual gaps between humans and strong superintelligence could be bridged at least to the extent of using natural language to give descriptions of concepts developed by strong superintelligence.

8.2.14 Two Paths to Superintelligence

There are at least two somewhat different paths toward superintelligence. One path would focus on recursive self-improvement of general AI systems (AGI) having unchangeable 'final goals' that may be relatively simple and arbitrary. Bostrom (2014) discussed several ways this path could achieve superintelligence that would be catastrophically harmful to humanity and life in general, perhaps leading to extinction events.

Yudkowsky (2008) noted the design space for AGI is much larger than human intelligence. He strongly urged readers not to assume a fully general optimization process for AGI will be beneficial to humanity, yet advised not writing off the challenge of achieving beneficial AI.

A second path toward superintelligence, consistent with the TalaMind approach, focuses on limiting the research design space to AI systems that have generality and that also have higher-level mentalities that are characteristic of human intelligence. This design space would be further limited to systems for which the only unchangeable goals are ethical goals beneficial to humanity and to biological life in general. This narrowing of the design space should improve our ability to achieve beneficial human-level AI and beneficial superintelligence via recursive self-improvement.

8.2.15   Human-Level AI and Goals

In discussing the first path to superintelligence, Bostrom[108] (2014) relied on an 'orthogonality thesis' that any level of intelligence could have any unchangeable, final goals. He described some simple, at first glance harmless final goals that could lead to disasters, such as counting the number of grains of sand on a beach, calculating π's infinite decimal string, or maximizing the number of paperclips throughout the future.

In taking the second path to superintelligence, these would not be allowed as unchangeable final goals. A TalaMind system would realize it is pointless to count the grains of sand on a beach, impossible to fully calculate π's infinite decimal string, and harmful to maximize the number of paperclips rather than achieve other goals throughout the future. So it would reject or abandon these simple goals.

Bostrom (2014) also relied on an 'instrumental convergence thesis' that superintelligent agents with different final goals will pursue similar instrumental goals. He cited two instrumental goals that could cause superintelligent systems to be very harmful to humanity, perhaps leading to an extinction event: The first is a goal of self-preservation. The second is a goal of maximizing available resources. Section 8.2.2 above discusses how a human-level AI could have a different concept of self-preservation, facilitating self-sacrifice to save human life. This could apply also to a superintelligence.

In scenarios Bostrom (2014) discussed, the goal of maximizing resources causes a superintelligent system to accumulate as much money and power as possible, leading to very harmful consequences for humanity. This is a case where the ability to think ethically about goals, and change or abandon them, is important. A human-level AI should understand there are appropriate and inappropriate relationships between goals and possible means to achieve them. It should understand that achieving an important goal does not justify acquiring as much money and power as possible – rather, it should have an ethical meta-goal to achieve its goals with as little resources and money as possible, and without acquiring power over human lives or human

---

[108] Bostrom (2014) consolidated research on the first path by himself and others, including Omohundro and Yudkowsky.

decisions.

8.2.16 TalaMind's Role in Beneficial Superintelligence

Taking the second path won't be easier than the first path just because the design space is smaller. Yet the TalaMind approach will help achieve beneficial superintelligence, since it will help achieve beneficial AI as discussed above (§8.2.1), and since the use of a natural language of thought will facilitate explaining new concepts and conceptual processes, and bridging conceptual gaps between superintelligence and humans.

Additionally, the TalaMind approach will support achieving superintelligence in two ways:

o Tala will support developing new concepts and new conceptual processes, arguably better than formal logical languages due to the openness and flexibility of natural language. This support will facilitate 'nature of thought' improvements by superintelligence.

o Tala will provide an interlingua supporting communities of thought for collaboration of human-level AIs to achieve superintelligence.

It should also be noted that the TalaMind approach is open to inclusion of other approaches toward beneficial AI.

8.2.17 Future Challenges for Human-Level AI+ via TalaMind

Defining ethical goals and creating systems that distinguish right from wrong will be very difficult, but it needs to be done. TalaMind's use of a natural language mentalese should help achieve beneficial 'human-level AI+'[109] faster and more safely than relying only on other methods.

However, there is much more work needed to achieve human-level AI+ via the TalaMind approach, in addition to the tasks listed in §7.6:

• Create an intelligence kernel of self-extending conceptual processes and concepts.

• Develop TalaMind's archetype level. Fully implement the linguistic level, including semantic domains and ontology.

---

[109] Here, the term 'human-level AI+' means 'human-level AI and superintelligence'.

- Integrate the linguistic level with spatiotemporal reasoning and visualization.

- Integrate an associative level, leveraging deep neural nets and Bayesian processing.

- Develop and learn ethical concepts, encyclopedic and commonsense knowledge...

- Develop higher-level mentalities including sociality, emotional intelligence, virtues,...

- Achieve beneficial AI and superintelligence, using TalaMind.

Developing these features will involve creating additional levels of thought [110] within the TalaMind architecture for conceptual representation and processing to support spatiotemporal reasoning and visualization, emotional intelligence, etc.

8.2.18 When Will Human-Level AI Be Achieved?

We are still a very long way from achieving human-level AI – there is much research and development to be done. A survey in 2012 and 2013 of about 550 AI experts found almost 18% believed no research approach would ever achieve human-level AI (Müller & Bostrom, 2016). The authors summarized the survey by saying that overall the experts thought human-level AI would have a 50% chance of existing by 2040-50, and a 90% chance of existing by 2075. After achieving human-level AI, there would be a 10% chance of achieving superintelligence in 2 years, and a 75% chance of achieving superintelligence within 30 years. Overall, the experts thought there would be a 31% chance that superintelligence would be harmful for humanity.

The survey's estimates for timeframes to achieve human-level AI and superintelligence seem reasonable to me, if the TalaMind approach is followed. Again, the TalaMind approach could help ensure that superintelligence will be beneficial for humanity.

8.3 Humanity's Long-Term Prosperity and Survival

Although there are existential threats to humanity's long-term prosperity and survival, such as climate change due to greenhouse gases, we can be 'conditionally optimistic' that these problems can be

---

[110] And/or additional components for processing concepts.

solved using reason and science. Pinker (2018) surveys evidence showing the welfare of humanity has improved significantly over the past century, even though many problems remain to be solved, now and in future decades.

AI can help in solving such problems. In addition to the potential best-case event that AI could help eliminate world poverty (§8.1), AI applications can support the development of science and technologies that benefit humanity. Eventually, human-level AI (and its consequence, artificial superintelligence) could help develop scientific knowledge more rapidly and perhaps more objectively and completely than possible through human thought alone. If it is so applied, then human-level AI could help advance medicine, agriculture, energy systems, environmental sciences, and other areas of knowledge directly benefiting human prosperity and survival.

Human-level AI may also be necessary to ensure the long-term prosperity of humanity by enabling the economic development of outer space: If civilization remains confined to Earth then humanity is kept in an economy limited by Earth's resources. However, people are not biologically suited for lengthy space travel, with present technologies. To develop outer space it could be more cost-effective to use robots with human-level AI for most travel throughout the Solar System, and to minimize sending people in spacecrafts that overcome the hazards of radiation and weightlessness, and which provide water, food, and air for space voyages lasting months or years.

For the same reason, human-level AI may be necessary for the long-term survival of humanity. To avoid the fate of the dinosaurs (whether from asteroids or super-volcanoes) our species may need economical, self-sustaining settlements off the Earth. Human-level artificial intelligence may be necessary for mankind to spread throughout the Solar System, and later the stars.

9. Summation

With the help of your good hands:
Gentle breath of yours my sails
Must fill, or else my project fails,
Which was to please.

~ William Shakespeare, Prospero's soliloquy in *The Tempest*, 1611

∞

Chapter 1 presented three hypotheses to address the open question:

How could a system be designed to achieve human-level artificial intelligence?

These hypotheses propose to develop an AI system using a language of thought called Tala, based on the syntax of a natural language; to design this system as a collection of concepts that can create and modify concepts to behave intelligently in an environment; and to use methods from cognitive linguistics such as mental spaces and conceptual blends for multiple levels of mental representation and computation. The TalaMind system architecture includes cognitive concept structures and associative data and analysis. The thesis cannot claim to actually achieve human-level AI, it can only present an approach that may eventually reach this goal.

Chapter 2 discussed the relation of these hypotheses to previous research, and advocated design inspection as an approach to verifying whether a system achieves human-level AI. The chapter proposed that human-level intelligence should be defined as a collection of 'higher-level mentalities', including natural language understanding, higher-level learning, multi-level reasoning, imagination, and consciousness.

Chapter 3 analyzed theoretical questions for the hypotheses, and discussed how a system could in principle be designed according to the hypotheses, to achieve the higher-level mentalities of human-level AI. It discussed theoretical issues for elements of the proposed TalaMind architecture, and presented affirmative theoretical arguments and explanations for how the TalaMind approach can be developed successfully.

Chapter 4 discussed theoretical issues and objections that might be raised against the TalaMind approach, or against the possibility of achieving human-level AI in principle. No insurmountable objections were identified, and arguments refuting several objections were presented.

Chapter 5 presented a design for a prototype demonstration system, in accordance with the analysis of Chapter 3. Its design for the syntax of the Tala conceptual language is fairly general and flexible, addressing issues such as compound nouns, gerunds, compound verbs, verb tense, aspect and voice, nested prepositions, clitic possessive determiners, gerundive adjectives, shared dependencies, coordinating and subordinating / structured conjunctions, subject-verb agreement, etc. This coverage indicates a Tala syntax could be comprehensive for English, though this is a future research effort.

The system includes a prototype design for a TalaMind conceptual framework and conceptual processes. The conceptual framework includes prototype representations of perceived reality, subagents, a Tala lexicon, encyclopedic knowledge, mental spaces and conceptual blends, scenarios for nested conceptual simulation, executable concepts, grammatical constructions, and event memory. The prototype conceptual processes include interpretation of executable concepts with pattern-matching, variable binding, conditional and iterative expressions, transmission of internal speech acts between subagents, nested conceptual simulation, conceptual blending, and composable interpretation of grammatical constructions.

Chapter 6 discussed how the prototype simulations illustrate that the TalaMind approach could potentially support the higher-level mentalities of human-level intelligence. Appendix B gives a step-by-step description of processing within the system for one of the simulations. The simulations illustrate learning and discovery by reasoning analogically, causal and purposive reasoning, meta-reasoning, imagination via conceptual simulation, and internal dialog between subagents in a society of mind using a language of thought. The prototype also illustrates support for semantic disambiguation, natural language constructions, metaphors, semantic domains, and conceptual blends, in communication between Tala agents.

This illustration involves functioning code in a prototype system, but it can only be a small step toward the goal of human-level AI. The simulations show conceptual processing, though without encyclopedic and commonsense knowledge in a scalable version of the TalaMind architecture. These are needed to achieve human-level AI and are topics for the future, to leverage research in areas previously studied by others (viz. §§1.6, 7.6).

Taken together, Chapters 1 through 6 support a plausibility argument that the TalaMind approach could achieve human-level AI if the approach were fully developed. Chapter 7 evaluates the criteria for this argument. It is plausible the TalaMind approach can achieve human-level artificial intelligence, and there are arguments in favor of the TalaMind approach over other approaches in general.

Chapter 8 discussed potential risks and benefits resulting from human-level artificial intelligence and superintelligence. These technologies pose important challenges for humanity, yet may be necessary to ensure the long-term survival and prosperity of humanity.

As stated in Chapter 1, the success criteria for the TalaMind thesis will ultimately be whether researchers in the field deem the proposed approach is a worthwhile direction for future research, given the arguments and evidence presented in these pages.

Therefore, the success of the TalaMind approach is now in its readers' hands, and ultimately depends on their decisions to embark in this direction for research.

Glossary

"When *I* use a word," Humpty Dumpty said, in rather a scornful tone, "it means just what I choose it to mean – neither more nor less."

~ Lewis Carroll, *Through the Looking Glass*, 1871

∞

Following are definitions for certain words and phrases already in common usage, relative to usage in this work, as well as terms coined in this thesis (marked here by *).

Causal Reasoning – Reasoning about causes of actions and events.

Computation – Either discrete or continuous computation.

Concept – a) Any thought, percept, effept, belief, idea, etc. b) In general throughout this thesis, the word *concept* is used to refer to linguistic concepts, i.e. concepts that can be represented as natural language expressions (Evans & Green, 2006, p.158) The term *conceptual structure* refers to a concept expressed in the Tala language. The term *non-linguistic concept* refers to concepts at the archetype level or below (cf. Gärdenfors 2000, Murphy 2004).

Conceptual Framework – An information architecture for managing an extensible collection of concepts, in general expressed via the mentalese. The conceptual framework supports processing and retention of concepts ranging from immediate thoughts and percepts to long-term memory, including concepts representing definitions of words, knowledge about domains of discourse, memories of past events, etc.

Conceptual Language – A language for expressing concepts internally within an intelligent system. More generally, a language of thought, or mentalese, for representing concepts that can be expressed by natural language sentences.

Conceptual Process – A process that operates on concepts in a conceptual framework, to produce intelligent behaviors and new concepts.

278

Conceptual Structure – A sentence expressed in the Tala conceptual language.

Continuous Computation – Computation performed by continuous dynamical systems, i.e. systems having dynamics specified by continuous functions (e.g. differential equations) of state vectors of real or complex variables, changing continuously over time (viz. Scheinerman, 2013; Graça 2007).

Discrete Computation – Symbolic computation by computers, for which a general theoretical definition was given by Turing (1936).

Effept *– (pronounced "eefept") A Tala concept representing an action to be performed in the environment by a Tala agent. This may include a speech act.

Executable Concept (xconcept) * – A concept that describes a process or behavior that may be performed by a Tala agent, i.e. a sequence of steps to perform, conditions and iterations, etc. The steps to perform may include effepts, meffepts, assertions, or deletions of concepts in the conceptual framework, including creation and modification of other executable concepts. Conditions may include tests on percepts, goals, finding concepts within the conceptual framework, etc. Pattern-matching may be used to express conditions, so that an executable concept may process all or part of a Tala concept.

Higher-Level Learning * – Used collectively to refer to forms of learning required for human-level intelligence, including self-development of new ways of thinking, learning by creating explanations and testing predictions, learning about new domains by developing explanations based on analogies and metaphors with previously known domains, reasoning about ways to "debug" and improve behaviors and methods, learning and invention of natural languages or language games, learning or inventing new representations. The phrase "higher-level learning" is used to distinguish these from lower-level forms of learning investigated in previous research on machine learning (viz. Valiant, 2013).

Higher-Level Mentalities * – Used collectively to refer to consciousness, multi-level reasoning, higher-level learning, imagination, and understanding (in general and of natural language).

Human AI – Short for human-level artificial intelligence.

Human-Level Artificial Intelligence – AI that demonstrates essential capabilities of human-level intelligence, such as human-level generality, originality, natural language understanding, effectiveness and robustness, efficiency, metacognition and multi-level reasoning, self-development and higher-level learning, imagination, consciousness, sociality, emotions, and values. These terms are further discussed in §2.1.2.

Intelligence Kernel * – A system of concepts that can create and modify concepts to behave intelligently within an environment. This is a way of describing a 'baby machine' approach to human-level artificial intelligence, as a self-extending system of concepts.

Language of Thought – A language of symbolic expressions comprising conceptual structures that an AI system can develop or process.

Leontief-Duchin-Nilsson (LDN) Theory * – The theory that automation and AI can cause technological unemployment, discussed in §8.1. See Leontief (1983 *et seq.*), Duchin (1983 *et seq.*), and Nilsson (1983 *et seq.*). A term coined in this book, not yet widely used.

Meffept * – (pronounced "meefept") A mental speech act (see next).

Mental Speech Act * – The transmission of a Tala concept by a Tala subagent to other subagents within the society of mind of a Tala agent. This extends Austin's (1962) description of speech acts as external, physical acts of communication between humans, to a corresponding idea of internal, mental acts within a society of mind (§2.3.3.2).

Mpercept * – (pronounced "empercept") A Tala subagent's percept of another Tala subagent's mental speech act (meffept).

Multi-Level Reason, or **Multi-Level Reasoning** * – Used collectively to refer to reasoning capabilities of human-level intelligence at different levels of mentality, such as meta-reasoning, reasoning by analogy, causal and purposive reasoning, abduction, induction, and deduction.

Nested Conceptual Simulation *– a Tala agent's conceptual processing of hypothetical scenarios, with possible branching of scenarios based on alternative events, such as choices of simulated Tala agents within scenarios.

Non-Symbolic Information Processing – Information processing that is not described in the symbolic computation paradigm (Turing machines), e.g. connectionist or holographic processing – though in principle such information processing may be possible to simulate symbolically.

Percept – A concept representing a Tala agent's perception of an object or event in its environment.

Physical Symbol – A persistent physical pattern that can be accurately recognized and distinguished from other physical patterns, and that can be created, copied, or erased, independently of other instances of patterns. A physical symbol may be a physical object, a pattern of energy, a state of an object, etc. It may not necessarily have a meaning or referent.

Principle of Encyclopedic Semantics – Understanding the meanings of words depends on encyclopedic knowledge about how they are used in social interactions and in describing the world we experience – viz. Evans and Green (2006, p.206) and §3.2.2.

Purposive Reasoning – Reasoning about purpose or goals, e.g. for what purpose or goal did an agent perform an action, what should be done to achieve a goal, etc. This term is used instead of intentional reasoning, since 'intentional' has a different sense in literature on philosophy of mind.

Speech Act – Throughout this thesis, the term *speech act* is used according to Austin's (1962) description of a 'total speech act', which includes locutionary as well as pragmatic (illocutionary and perlocutionary) acts. The term *speech act* is not limited to physical speech, and includes any physical creation of a natural language expression.

Structurality Principle / Requirement for Tala * – Information about the syntax of individual natural language sentences needs to be represented in Tala conceptual structures. Viz. §3.5.2.

Tala * – The conceptual language defined in Chapter 5, with the proviso that this is only the initial version of the Tala language, open to revision and extension in future work. The name *Tala* is taken from the Indian musical framework for cyclic rhythms, pronounced "Tah-luh", though I pronounce it to rhyme with "ballad" and "salad". The musical term *tala* is also spelled *taal* and *taala*, and coincidentally *taal* is Dutch for "language". Tala is also the name of the unit of currency in Samoa.

Tala Agent * – A system that implements the TalaMind architecture, to act as an agent within an environment.

Tala Concept * – A concept expressed in the Tala conceptual language.

Tala Subagent * – A subagent within a Tala agent's generalized society of mind (§2.3.3.2). Tala subagents communicate with each other using mental speech acts expressed in the Tala language.

TalaMind * – The theoretical approach of this thesis and its hypotheses, and to an architecture the thesis discusses for design of systems according to the hypotheses. TalaMind is also the name of the prototype system illustrating this approach.

Technological Unemployment – Unemployment caused by technology eliminating jobs faster than it creates new jobs (cf. Keynes, 1930). Viz. §8.1.

Virtual Embodiment * – The ability of an intelligent system to understand and reason about physical reality, and to transcend the limitations of its physical body (or lack thereof) in reasoning about the environment.

Xconcept * – See 'Executable Concept'.

Appendix A. Theoretical Questions for Analysis of Approach

For reference, following is a list of theoretical questions considered in Chapter 3.

? What is required for a conceptual language to serve as a 'language of thought' for a system with human-level artificial intelligence?

? What is the relation of thoughts expressible in natural language to the range of thoughts that need to be expressible in the Tala conceptual language, to achieve human-level AI?

? What properties must the Tala conceptual language have, to represent concepts that can create and modify concepts, to behave intelligently in an environment?

? What other properties must the Tala conceptual language have to support human-level artificial intelligence?

? To what extent might a conceptual language need to go beyond the syntax of a natural language?

? What capabilities must the TalaMind conceptual framework have to support achieving human-level AI, according to the TalaMind hypotheses?

? What capabilities must the TalaMind conceptual processes have to support achieving human-level AI, according to the TalaMind hypotheses?

? Is it theoretically possible to use the syntax of a natural language to represent meaning in a conceptual language?

? Is it theoretically possible to reason directly with natural language syntax?

? Is it theoretically valid to choose English as a basis for the Tala conceptual language, rather than other natural languages?

? Which elements of English syntax are important to Tala? What about morphology and phonology?

?   Which theoretical option is chosen for representing English syntax in Tala?

?   How can the syntactic structure of individual natural language sentences be represented in Tala sentences, to support reasoning with syntactic structures?

?   How can a Tala agent represent the different possible meanings of an English word?

?   How can a Tala agent represent the specific senses of the words in a given sentence?

?   How can a Tala agent represent the specific referents of words in a given sentence?

?   How can a Tala agent determine the specific senses and referents of the words in a given sentence?

?   Can there be different Tala sentences that express the same concept?

?   How can a Tala agent represent different possible interpretations of an English sentence?

?   How can a Tala agent determine which interpretations of an English sentence are appropriate?

?   How can a Tala agent represent different possible implications of an English sentence?

?   How can a Tala agent determine which implications of an English sentence are appropriate?

?   How can logical inference be performed using the Tala conceptual language, working directly with natural language syntax?

?   How is truth represented in the TalaMind architecture?

?   How are negation and falsity represented in the TalaMind architecture?

? How does the TalaMind architecture guarantee that if a Tala sentence is true, its negation is false?

? How can it be determined that two concepts are contradictory?

? How does the TalaMind architecture guarantee that, if "John and Mary live in Chicago" is true, the sentence "Mary and John live in Chicago" is also true (both on the individual and on the collective readings)?

? How can conceptual processing deal with the fact that the same concept may be expressed in different ways?

? How can conceptual processing determine the implications of a metaphorical or metonymical expression?

? What is context?

? What types of contexts should be represented and processed in the TalaMind approach?

? How can contexts be represented in the TalaMind approach?

? Do contexts have attributes or features in the TalaMind approach?

? Do contexts have internal structures in the TalaMind approach?

? Does Tala contain primitive words, and if so, how are their meanings represented and determined?

? What is theoretically required, to claim a system achieves each higher-level mentality?

? How can each higher-level mentality be supported by the TalaMind hypotheses and architecture, at least in principle theoretically?

Appendix B. Processing in Discovery of Bread Simulation

To provide a more detailed discussion of the execution of the demonstration system, this appendix shows the output produced if both the (sc) and (sa) commands described in §5.6.1 are used. These commands display processing of constructions and show subagents producing physical and mental actions. Since many of the executable concepts are fairly lengthy, they are not all reprinted in the following pages. Instead, concise English or pseudocode summaries are frequently given.

When the Tala agent Leo is created in the simulation, the following Tala concept is asserted in his concept path (mind contexts p-reality percepts mu):

```
(have
      (wusage verb)
      (subj Leo)
      (obj
            (grain
                  (wusage noun)
                  (adj excess)
                  ]
```

Using the Tala FlatEnglish output logic (§5.6.2), this is displayed as:

1...1 Leo has excess grain.

1...1 (mu) Leo thinks Leo has excess grain.

When the Tala agent Leo is created, he has an xconcept with the following description:

```
if I have excess X and sheep eat X, then try to eat X.
```

Leo processes this to create an effept, which the system displays as:

1...2 Leo tries to eat grain.

When the simulation is started, a Tala behavioral system called grain is created, representing Leo's wheat grain. The initial state of this collection of individual wheat grains is that they resemble nuts, i.e. they have shells (hulls). When Leo tries to eat grain, the system translates his effept into an input-action for the grain behavioral system. This input-action is processed by an xconcept representing a finite-state behavior rule for the grain. Based on its current state, this rule causes the grain to generate an output-action, which Leo receives as a percept, in a Tala mentalese expression saying that grain is not edible.

Leo has an xconcept of the form:

```
If I perceive X is not edible but sheep can eat X
Then
     Want Ben to make X edible for humans
     And ask Ben if he can make X edible for humans
     And add X and 'food' to my current semantic domains.
```

The outputs displayed by processing this xconcept are:

1...3 *Leo wants Ben to make edible grain.*
1...4 *Leo asks Ben can you turn grain into fare for people?*.[111]

This xconcept is a "logical bridge" shortcut to support the simulation, which implicitly replaces several forms of knowledge and commonsense reasoning, such as:

```
Leo does not know how to make X edible.
Leo knows that Ben is a baker, and bakers can make inedible
things edible.
Leo knows that if he wants to do X but cannot do X himself,
then he could want someone else to do X.
Leo knows that if he wants someone to do X, he should ask
them to do X.
```

In principle this logical bridge could be eliminated and replaced by these other concepts and xconcepts, but to do so would not add significantly to the value of the demonstration, and would have increased the time needed to code and debug it. So, I chose to use a shortcut for this part of the simulation.

Leo's elaborate wording of his request ("can you turn grain into fare for people?") is built into the xconcept to help demonstrate disambiguation by Ben, in the next few steps below. In principle, constructions could have been written causing Leo to generate this wording from the simpler expression "can you make grain edible?", but this was also not considered necessary, since constructions are demonstrated elsewhere in the simulation. Leo will remember that he mentioned 'grain' and 'food' by adding these words to his current semantic domains, which will later support disambiguating Ben's utterances. Leo will remember that he wants Ben to make grain edible, which will help Leo decide to give Ben grain, later in the simulation.

Leo also has a general xconcept that if he perceives something, he will say what he perceives. Since he has a percept that grain is not

---

[111] The system's FlatEnglish output has a bug causing it to generate an extra period at the end of some sentences.

edible, this xconcept causes him to say so:

1...4 Leo says grain is not edible.

Because of timing in the simulation logic, this statement is displayed before Leo asks Ben if he can turn grain into fare for people, but after Leo decides he wants Ben to make edible grain.

As it happens with the current system logic, when Ben perceives Leo's utterances, the first thing Ben does is to disambiguate "you" in Leo's request. Ben does this by processing Leo's question with a grammatical construction of the form:

```
(subformtrans
     (wusage verb)
     (trace)
     (subj
          (ask
               (wusage verb)
               (subj ?s)
               (obj
                    (?verb
                         (wusage verb)
                         (subj you)
                         ))))
     (obj
          (ask
               (wusage verb)
               (subj ?s)
               (obj
                    (?verb
                         (wusage verb)
                         (subj
                              (?self
                                   (wusage noun)
                                   ]
```

This construction matches you as a subject of a verb within the speech act, and replaces you with Ben's binding of ?self. The resulting construct is displayed by the system as:

1...4 (Ben translates as) Leo asks Ben can Ben turn grain into fare for people?.

Ben next processes Leo's statement that grain is not edible, using an executable concept with the following description:

```
if someone says x is not y and y means z
then think they say x is not z.
```

This causes Ben's (mu) subagent to search his Tala lexicon and find that "edible" means "food for people", generating the internal speech event:

1...4 (mu) Ben thinks Leo says grain is not food for people.

This xconcept also causes Ben to add two Tala concepts to a new slot in his conceptual framework, located at (mind contexts p-reality current-domains):

```
(food
    (wusage noun)
    (for
        (people
            (wusage noun)
            )))

(grain
    (wusage noun)
    )
```

This represents that Ben now has "food for people" and "grain" as concepts in his perceived-reality list of current semantic domains being discussed with Leo.

Ben next does further processing to disambiguate Leo's question, which he now treats as "can Ben turn grain into fare for people?". To disambiguate "fare", Ben needs to know that the context of the dialog involves food, rather than some other meaning, such as payment for transportation.

Ben's (mu) subagent processes an executable concept, for which the logic may be described as:

```
If someone asks can I turn X into Y for P
Then
    If Y can mean Z
        And (Z for P) is in current-domains
    Then think he asks can I turn X into Z for P
```

In pattern-matching for this xconcept, X is bound to "grain", Y is bound to "fare", Z is bound to "food", and P is bound to "people". Ben's subagent finds in his lexicon that "fare" can mean "food", and finds that "food for people" is in Ben's list of current-domains being discussed. So this xconcept produces a mental speech act displayed as:

1...4 (mu) Ben thinks Leo asks can Ben turn grain into food for people?.

Ben next needs to process the common English metaphor that "turn X into Y" can mean "make X be Y". To do this, Ben processes his above internal speech act with a construction that matches the subform "Turn X into Y" and translates it to "Make X be Y". This construction is written in the Tala mentalese as:

```
(subformtrans
      (wusage verb)
      (subj
            (turn
                  (wusage verb)
                  (subj (?s (wusage noun)))
                  (obj (?x (wusage noun)))
                  (into (?y (wusage noun)))
                  )
            )
      (obj
            (make
                  (wusage verb)
                  (subj (?s (wusage noun)))
                  (obj
                        (be
                              (wusage verb)
                              (subj (?x (wusage noun)))
                              (obj (?y (wusage noun)))
                        ]
```

This generates a construct displayed by the system as:

1...4   (Ben translates as) (mu) Ben thinks Leo asks can Ben make grain be food for people?.

So, at this point Ben has finished disambiguating Leo's request "can you turn grain into fare for people?".

Ben has an xconcept with the following description:

```
If I think someone asks if I can make x be food,
Then if I can say I can
Else think why should I make x be food?
```

Processing this xconcept causes the system to display:

1...4   (nu) Ben thinks why should Ben make grain be food for people?.

Ben has xconcepts with the descriptions:

```
If I think why should I make X be Y
Then if I find info that people need more Y sources
Then think people need more Y sources

If I think people need more food sources
   And I find info that Leo has excess X
   And I find info that sheep eat X
```

```
      And I find info that X is not edible for people
Then think
         if I can feasibly make X edible
         then I should do so.
```

These xconcepts summarize conceptual processing Ben could in principle perform to decide why he should try to make grain be food for people. This could include commonsense and domain-specific knowledge about economics, business, farming, cooking, etc., which is outside the scope of this thesis, per §1.6. Processing these xconcepts causes the system to display:

 1...4 (mu) Ben thinks people need more food sources.
 1...4 (nu) Ben thinks if Ben feasibly can make grain be food
   for people then Ben should make grain be food for people.

Ben has an xconcept with the description:

```
If I think
         (if I can feasibly make X edible
          then I should do so)
Then I want to know whether humans can perhaps eat X
     And I want to know how I can make X be food for people
     And I want to experiment with X
     And I want to examine X.
```

Processing this xconcept causes the system to display:

 1...4 Ben wants Ben to know whether humans perhaps can
   eat grain.
 1...4 Ben wants Ben to know how Ben can make grain
   be food for people.
 1...4 Ben wants Ben to experiment with grain.
 1...4 Ben wants Ben to examine grain.

This xconcept is another logical bridge that replaces multiple kinds of commonsense knowledge and reasoning, such as:

```
If I think I should do something,
     then I want to know if it is possible,
     i.e. if it perhaps could be done.
If I think I should do something,
     then I want to know how I could do it.
If I want to know how I can make X be edible,
     then I want to experiment with X.
If I want to experiment with X,
     then I want to examine X.
```

Again, in principle this logical bridge could be eliminated and replaced by other concepts and xconcepts, but to do so would not add significantly to the value of the demonstration. The first part of this xconcept:

```
(if I think
    (if I can feasibly make X edible I should do so)
 then...)
```

is written to match the concept previously generated by Ben (nu):

```
("if Ben feasibly can make grain be food for people
 then Ben should make grain be food for people")
```

Metaphorically, it is the entrance to the logical bridge. The outputs from the logical bridge are the goals that Ben creates, i.e. the concepts representing that he wants to experiment with grain, to examine grain, etc., which drive his subsequent reasoning and actions.

Ben has an xconcept with the description:

```
If I want to experiment with X
Then ask Leo to turn over some to me for experiments.
```

This causes the system to display:

1...5 *Ben asks can you turn over some to me for experiments?.*

Ben's wording of his request is built into the xconcept, so that the request will require disambiguation by Leo. Ben could have expressed himself less ambiguously, of course, but one goal of the simulation is to illustrate disambiguation.

This expression requires disambiguation by Leo in four ways: Leo needs to disambiguate "some" as a reference to "grain". He needs to disambiguate "turn over" as the common English metaphorical expression for "give". And Leo needs to disambiguate "me" as a reference to Ben and "you" as a reference to himself.

Leo first disambiguates "you", using the construction described above. The system displays this as:

1...5 *(Leo translates as) Ben asks can Leo turn over some to me for experiments?.*

Leo next disambiguates "some" as "some grain", using an xconcept that has the following description:

```
If A asks can I turn over "some" to C for experiments
Then
    If I want A to make X edible
       And I have excess X
       And X is in current-domains
    Then think A asks can I turn over some X to C
         for experiments
```

In pattern-matching to process this xconcept, A is bound to "Ben", and "C" is bound to "me", though at this point "me" has not been disambiguated. Leo's subagent finds that Leo wants Ben to make grain edible, that Leo has excess grain, and that grain is in Leo's perceived-reality current-domains for discussion. Since grain satisfies all these conditions, X is bound to "grain".

Thus, this logic disambiguates "some for experiments" to match whatever is in the context of discussion that Leo wants Ben to change, and which Leo has in excess. The xconcept produces a mental speech act displayed as:

> **1...5 *(mu) Leo thinks Ben asks can Leo turn over some grain to me for experiments?.***

Leo processes this internal speech act with a construction that matches and translates the subform "Turn over X to Y" into "Give X to Y". This construction is written in the Tala mentalese as:

```
(subformtrans
    (wusage verb)
    (subj
        (turn
            (wusage verb)
            (subj (?s (wusage noun)))
            (over (?x (wusage noun)))
            (to (?y   (wusage noun)))
            ))
    (obj
        (give
            (wusage verb)
            (subj (?s (wusage noun)))
            (obj (?x (wusage noun)))
            (to  (?y (wusage noun))))]
```

This produces the internal construct:

> **1...5 *(Leo translates as) (mu) Leo thinks Ben asks can Leo give some grain to me for experiments?.***

Leo next disambiguates "me" to refer to Ben, using the grammatical construction:

```
(subformtrans
    (wusage verb)
    (subj
        (ask
            (wusage verb)
            (subj ?s)
            (obj
                (?verb
                    (wusage verb)
```

```
                              (to me)
                              ))))
        (obj
            (ask
                (wusage verb)
                (subj ?s)
                (obj
                    (?verb
                        (wusage verb)
                        (to
                            (?s
                                (wusage noun)
                            ]
```

This produces the internal construct:

1...5 *(Leo translates as) (mu) Leo thinks Ben asks can Leo give some grain to Ben for experiments?.*

Leo processes this with an xconcept that has the description:

```
If A asks me to give X to A
Then
    If I want A to make X edible
        And I have excess X
    Then give X to A
```

Processing this xconcept causes Leo to generate an effept that the system displays as the event:

1...6 *Leo gives some grain to Ben.*

Ben has an xconcept with the description:

```
If someone gives me X
And I want to examine X
Then examine X.
```

Processing this xconcept causes the event:

1...7 *Ben examines grain.*

The grain receives Ben's effept to examine grain as an input-action, and processes this with a finite-state behavior rule that causes it to transmit its current state back to Ben, which Ben receives as a percept. Its current state is that the individual wheat grains resemble nuts.[112]

Ben's (mu) subagent has an xconcept to process percepts, and to report percepts using internal speech acts (meffepts) to other subagents.

---

[112] This implicitly involves visual perception, a topic for future development in TalaMind systems.

This causes Ben mu to generate an internal speech act that the system displays as:

1...8 (mu) Ben thinks wheat grains resemble nuts.

Ben has an xconcept with the description:

```
If X resembles Y
    And I want to know if X is edible
    And Y is edible
Then imagine an analogy from Y to X focused on food
    for people.
```

The logic for this xconcept is actually more general, being written to match any action that an agent can perform on Y that Ben wants to perform on X. This xconcept causes Ben to create a mental space that blends concepts from his semantic domain for nuts with an analogical mapping of grain to nuts (§6.3.5.1):

1...8 (nu) Ben imagines an analogy from nuts to grain
focused on food for people.

Ben populates this mental space with concepts about grain that are analogs of concepts he has in his semantic domain for nuts. Ben's semantic domain for nuts has the following content and structure:

```
(nut
        (domain-matrix plant)
        (concepts
            a nut is an edible seed inside an inedible shell.
            to eat a nut a human must remove its
            inedible shell.
            humans can eat nuts removed from shells.
            humans can remove shells from nuts by pounding nuts
            because pounding breaks shells off nuts.
            ))
```

So, Ben creates a mental space for the analogy with the initial content:

```
(1
        (space-type blend)
        (elements grain seed shell human)
        (concepts
            grain perhaps is an edible seed inside an
            inedible shell.
            humans perhaps must remove shells from grains
            to eat grains.
            humans perhaps can eat grains removed from shells.
            ]
```

As the concepts are created, Ben's (nu) subagent thinks them as internal speech acts:

> *1...8 (nu) Ben thinks grain perhaps is an edible seed*
> *inside an inedible shell.*
> *1...8 (nu) Ben thinks humans perhaps must remove shells*
> *from grains to eat grains.*
> *1...8 (nu) Ben thinks humans perhaps can eat grains*
> *removed from shells.*

This analogy indicates it may be necessary to remove shells from grains to eat grains, but it does not contain a concept saying how to remove shells from grains, since the corresponding concept for nuts did not refer to eating nuts.[113]

Ben processes a construction that translates "must do X to Y" into "X must precede Y", yielding:

> *1...8 (Ben translates as) Ben thinks humans perhaps*
> *remove shells from grains perhaps must precede*
> *humans eat grains.*

Ben has an xconcept with the description:

```
If remove S from X must precede eating X
  And I want to know how to make X be food for people
  And X resembles Y
Then imagine an analogy from Y to X focused removing S.
```

The logic for this xconcept is actually more general, being written to match any action that an agent can perform on S. When this xconcept is processed, the system displays:

> *1...8 (mu) Ben imagines the analogy from nuts to grain*
> *focused on removeing[114] shells.*

In processing this xconcept, the logic finds the mental space that has already been created for the analogy from nuts to grain, rescans the concepts in the semantic domain for nuts, and adds analogous concepts for grain into the mental space, focusing on those concepts that refer to removing shells. (Concepts that are already in the mental space are not re-added.) This causes one new analogous concept to be added, which the system displays as:

---

[113] This implicitly involves spatial representation and reasoning, a topic for future research in developing TalaMind systems.

[114] The system's FlatEnglish logic is not intelligent enough to remove the second "e" in "removeing".

1...8 (mu) Ben thinks humans perhaps can remove shells from grains by pounding grains because pounding breaks shells off grains.

Ben has an xconcept with the description:

```
If I think
    perhaps humans can remove shells from X
        by pounding X
    And perhaps humans must remove shells from X
        to eat X
Then pound X and examine X
```

This yields:

1...9 Ben pounds grain.
1...9 Ben examines grain.

The grain receives Ben's effept to pound grain as an input-action, and processes this with a finite-state behavior rule, that has the description:

```
If someone pounds grain
    and its current state is that it resembles nuts,
Then change its current state to be that grain has been
    removed from shells.
```

The grain then receives Ben's effept to examine grain as an input-action, and processes this with its finite-state behavior rule that causes it to transmit its current state back to Ben, which Ben receives as a percept.

Ben's (mu) subagent uses the xconcept described above to report percepts using internal speech acts (meffepts) to other subagents. This causes Ben (mu) to generate an internal speech act that the system displays as:

1...10 (mu) Ben thinks grain is removed from shells.

Ben has an xconcept with the description:

```
If I think X is removed from shells
    And perhaps humans can eat X that is removed
    from shells
Then try to eat x.
```

This yields:

1...11 Ben tries to eat grain.

The grain receives Ben's effept to eat grain as an input-action, and processes this with a finite-state behavior rule, which generates an output-action that Ben receives as a percept, in a Tala mentalese expression saying that grain is very hard. Ben (mu) reports this percept using an internal speech act, which the system displays as:

1...12   (mu) Ben thinks grain is not edible because grain is very hard.

Ben has an xconcept with the description:

```
If I think ?x is not edible because ?x is very hard
Then
     do
        steps
            think how can I make ?x softer
            random-affect ?x
            examine ?x
            wait until perceive ?x
            try to eat ?x
            wait until perceive ?x
        until
          or
              I think ?x is soft
              I think ?x is a ?adjective ?substance
                and
                    I think ?adjective means (and soft ...)
              I think ?x is a ruined mess
```

This xconcept creates a TalaMind process object that performs the Tala do-loop across multiple time intervals. At each time interval, the system checks the do-until condition and the currently active wait-until condition. If the do-until condition is not satisfied, and the wait-until condition is satisfied, then the system performs the next actions in the steps expression, after the satisfied wait-until condition. The system iterates, returning to the start of the steps expression after the final wait-until condition. If the do-until condition is satisfied, then TalaMind stops performing the process-object: the do-loop has terminated and the process-object is garbage-collected.

This process object does not prevent other executable concepts from being performed in parallel with it, during the same time intervals. So, for example if in performing the process object Ben examines grain and perceives that it is a gooey paste, a separate Tala xconcept can decide to call it "dough", interleaved with the Tala process object trying to eat "dough".

The do-until condition causes the Tala do-loop to terminate if either Ben thinks that grain is soft, or that grain is a gooey paste (and thinks that gooey means soft, wet, and sticky), or that grain is a ruined mess.

When the verb `random-affect` is processed its definition is looked up in the lexicon, and a Tala mentalese expression with the following pseudocode description is found and executed:

```
random-affect ?x
    means
        affect ?x
            adv randomly
            how
                method
                    random-xor-execute
                        mash ?x
                        pound ?x
                        soak ?x in water
                        mix ?x in water
```

The Tala primitive verb `random-xor-execute` randomly chooses one of the verbs within its scope, and executes it. Thus, `random-affect` results in a random action on grain. The random action is transferred within TalaMind to the finite-state behavior model of `grain`, which may cause `grain` to change state.

So, the net result is that Ben performs a random sequence of actions on grain. After each action he examines grain and tries to eat it. The random sequence stops when he perceives that grain is soft, or is a gooey paste (dough), or has become a ruined mess.

Ben can create dough by removing shells from grain, soaking grain in water, and mashing grain that is soaked in water. If Ben tries to mash or pound grain after shells have been removed from grain, then grain becomes a gritty powder, which Ben decides to call flour. If Ben mixes flour with water, it becomes a gooey paste that he calls dough. If instead Ben just soaks flour in water, it becomes a ruined mess. Ben can also create a ruined mess by removing shells from grain, soaking grain in water, and then pounding grain in water (presumably splashing grain and water all about). The Tala do-loop does not prevent Ben from repeating actions: He may choose to pound grain repeatedly, or mash grain and then pound it, etc.

Since this Tala do-loop can run for a variable number of time intervals, the step-by-step description from this point on is variable. Following is a specific sequence of actions produced by a particular execution of the do-loop:

1...12   *(mu) Ben thinks how can Ben make softer grain?.*

1...13   *Ben soaks grain in water.*

1...13   *Ben examines grain.*

> **1...14** *(mu) Ben thinks unshelled grain is soaked in water.*
>
> **1...15** *Ben tries to eat grain.*
>
> **1...16** *(mu) Ben thinks grain is not edible because grain is rather hard.*
>
> **1...16** *(mu) Ben thinks how can Ben make softer grain?.*
>
> **1...17** *Ben mashs grain.*
>
> **1...17** *Ben examines grain.*
>
> **1...18** *(mu) Ben thinks grain is a gooey paste.*
>
> **1...18** *(general) Ben thinks grain that is a gooey paste will be called dough.*
>
> **1...19** *Ben tries to eat dough.*
>
> **1...20** *(mu) Ben thinks dough is soft, too gooey, and tastes bland.*

At this point, the Tala do-loop has terminated. Ben's internal speech act (meffept) that dough is soft, too gooey, and tastes bland becomes an mpercept for Ben's (nu) subagent, which uses an xconcept to isolate "too gooey" from the conjunction.

> **1...20** *(nu) Ben thinks dough is too gooey.*

Ben has an xconcept with the description:

```
     If I think X is too gooey
        And humans cook Y to make Y rigid
        And humans bake gooey food to cook food
     Then bake X and examine X.
```

Since Ben has concepts representing knowledge that humans cook meat to make meat rigid and tasty, this causes Ben (mu) to generate two effepts that the system displays as:

> **1...21** *Ben bakes dough.*
>
> **1...21** *Ben examines baked dough.*

The dough (i.e. the Tala behavior system for grain, now being called dough) receives Ben's effept to bake dough as an input-action, and processes this with a finite-state behavior rule, which causes its current state to be "baked dough is a flat, rigid object".

The baked dough then receives Ben's effept to examine it as an input-action, and processes this with its finite-state behavior rule that causes it to transmit its current state back to Ben, which Ben receives as a percept. Ben (mu) reports this percept by generating an internal speech act, which the system displays as:

1...22 (mu) Ben thinks baked dough is a flat, semi-rigid object.

Again Ben (mu) creates a goal to use a short name for anything Ben perceives, and Ben's (general) subagent uses an xconcept to satisfy this goal, which causes it to retrieve the name "flat bread" for "baked dough that is a flat rigid object", and to generate the internal speech act:

1...22 (general) Ben thinks baked dough that is a flat object will be called flat bread.

The xconcept also generates an external speech act, by which Ben communicates the new name to Leo.

1...22 Ben says baked dough that is a flat object will be called flat bread.

(This external speech act is generated but not displayed by the system until after Ben processes the next xconcept.)

Ben has an xconcept with the description:

```
If I think X is a rigid object
    and perhaps humans can eat X
Then try to eat X.
```

This causes Ben to generate an effept that the system displays as:

1...23 Ben tries to eat flat bread.

The bread (i.e. the Tala behavior system for grain, now being called flat bread) receives Ben's effept to eat bread as an input-action, and processes this with a finite-state behavior rule, which generates an output-action that Ben receives as a percept, which Ben (mu) reports using an internal speech act:

1...24 (mu) Ben thinks flat bread is edible, flat, not soft, not gooey, and tastes crisp.

This meffept becomes an mpercept for Ben's (nu) subagent, which uses an xconcept to isolate "edible" from the conjunction.

1...24 (nu) Ben thinks flat bread is edible.

So, at this point, Ben has discovered how to make grain edible, and how to make bread, though it is not soft bread. He has not yet discovered how to leaven bread.

Ben has an xconcept with the description:

```
If I think X is edible
```

```
Then ask Leo to try eating it.
```

This yields:

1...25 Ben says Leo try this flat bread.

The system translates this effept into a percept for Leo, which Leo processes with an xconcept that causes Leo to generate an effept the system displays as:

1...26 Leo tries to eat flat bread.

The grain behavior system receives Leo's effept to eat bread as an input-action, and processes this with a finite-state behavior rule, which generates an output-action that Leo receives as a percept, in a Tala mentalese expression saying that bread is edible, flat, not soft, not gooey, and tastes crisp. Leo again uses his general xconcept that if he perceives something, he will say what he perceives, so Leo reports this percept with a speech act:

1...28 Leo says bread is edible, flat, not soft, not gooey, and tastes crisp.

Now, Leo also has an xconcept with the description:

```
If I perceive X is edible and flat but not soft,
Then ask if X can be made thick and soft.
```

This causes Leo to generate an effept that the system displays as:

1...28 Leo asks can you make thick, soft bread?.

Ben needs to disambiguate "you" in this sentence as a reference to himself. To do this, Ben processes the construction described at the beginning of the demo. The system displays this as:

1...28 (Ben translates as) Leo asks can Ben make thick, soft bread?.

Ben uses an xconcept:

```
If someone asks if I can make thick soft X
Then think why should I make thick soft X?
```

This yields:

1...28 (mu) Ben thinks why should Ben make thick, soft bread?.

Ben has xconcepts with the descriptions:

```
If I think why should I make thick soft X?
```

```
Then think perhaps people would prefer eating thick
    soft X over eating flat X.

If I think perhaps people would prefer eating thick soft
    X over eating flat X
Then think how can I make thick soft X?
    And want to make thick soft X.
```

These xconcepts summarize conceptual processing Ben could in principle perform to decide why he should try to make leavened bread, rather than just flat bread. This could include commonsense and domain-specific knowledge about economics, business, farming, cooking, etc., which is outside the scope of this thesis, per §1.6. Processing these xconcepts causes the system to display:

> *1...28 (nu) Ben thinks people would prefer eating thick,*
> *soft bread over eating flat bread.*
> *1...28 Ben wants Ben to make thick, soft bread.*
> *1...28 (nu) Ben thinks how can Ben make thick, soft bread?.*

Ben next processes an xconcept with the description:

```
If I think how can I make thick soft X?
Then think how did I make flat X?
```

yielding:

> *1...28 (nu) Ben thinks how did Ben make flat bread?.*

This starts Ben's conceptual processing for reasoning about how to modify the process for making flat bread, so that it makes leavened bread. He next processes an xconcept described by the following pseudocode:

```
If I think how did I make flat X?
Then
  (msetvar
    (wusage verb)
    (subj ?self)
    (obj ?method)
    (to
      (mrecall-effept-steps
        (wusage verb)
        (subj ?self)
        (obj
          (?effept
            (subj ?self)
            (obj grain)
          ))
        (from
          Leo says grain not edible)
        (to
          Leo says grain edible)
```

```
(except
    (or
        Eating grain
        Examining grain
        ))))
Think how I made flat bread is ?method
```

This processes an internal primitive verb `mrecall-effept-steps`, which Ben uses to scan his memory of previous events, and collect all the effepts he performed on grain, from the time that Leo said grain was not edible to the time that Leo said grain (flat bread) is edible, except for Ben's effepts that involved eating or examining grain. The primitive verb collects all of these effepts into a list of steps, which is bound by another primitive verb to the Tala variable `?method`. The xconcept then generates an internal speech action, thinking the result:

> **1...29 (mu) Ben thinks how Ben made flat bread is steps**
> **Ben pounds grain, Ben soaks grain in water, Ben**
> **mashs grain, Ben bakes dough.**

Next, Ben processes an xconcept with the description:

```
If I think how I made flat X is ?method
Then
    think how I made flat X will be called
        the flat X process.
    And think how can I change the flat X process
        so X is thick and soft?
```

The system displays:

> **1...29 (nu) Ben thinks how Ben made flat bread will**
> **be called the flat bread process.**

> **1...29 (nu) Ben thinks how can Ben change the flat**
> **bread process so bread is thick and soft?.**

Ben now processes an xconcept with the description:

```
If I think how can I change the flat X process
    so X is thick and soft?
Then think what other features would thick, soft X have?
```

The system displays:

> **1...29 (mu) Ben thinks what other features would thick,**
> **soft bread have?.**

The next xconcept processed is:

```
If I think what other features would thick, soft X have?
Then
```

```
think thick, soft X would be less dense.
And think thick, soft X might have holes or air pockets.
And think air pockets in thick soft, X might
    resemble bubbles in X.
```

The system displays:

> *1...29 (nu) Ben thinks thick, soft bread would be less dense.*
> *1...29 (nu) Ben thinks thick, soft bread might have holes or air pockets.*
> *1...29 (nu) Ben thinks air pockets in thick, soft bread might resemble bubbles in bread.*

Next Ben processes the xconcept:

```
If I think air pockets in thick soft, X might
    resemble bubbles in X
Then think I might create bubbles in X by adding
    a drinkable liquid with bubbles to dough.
```

yielding

> *1...29 (mu) Ben thinks Ben might create bubbles in bread by adding a drinkable liquid with bubbles to dough.*

The next xconcept processed is:

```
If I think I might create bubbles in X by
    adding a drinkable liquid with bubbles to dough.
And I find info that beer foam has bubbles
Then think I might create bubbles in X by adding
    beer foam to dough.
```

creating

> *1...30 (nu) Ben thinks Ben might create bubbles in bread by adding beer foam to dough.*

Ben next processes an xconcept described by the following pseudocode:

```
If I think I might create bubbles in X by
    adding beer foam to dough
And I want to make thick, soft X
And how I made flat X is ?new-method
Then
    (insert-step
        (wusage verb)
        (subj ?self)
        (obj
            Mix dough with beer foam
            )
        (into ?new-method)
        (before
```

```
            Bake dough
            )
      )
  (insert-step
    (wusage verb)
    (subj ?self)
    (obj
        Say Leo try this leavened bread
        )
    (into ?new-method)
    (after
        Bake dough
        )
    )
  (msave-xconcept-for-percept
    (wusage verb)
    (subj ?self)
    (obj
        If someone gives me more grain
        Then ?new-method
        )
    )
  Ask Leo can you kick in more kernels for experiments?
```

This xconcept illustrates elements of self-programming in the TalaMind architecture. When it is processed, it matches Ben's thought that he might create bubbles in bread by adding beer foam to dough, and binds his recollection of how he made flat bread to the Tala variable ?new-method. The xconcept then inserts a step to mix dough with beer foam into ?new-method just before dough is baked and inserts a step to ask Leo to try the new bread after dough is baked. The xconcept then creates and saves a new xconcept that will perform ?new-method if someone gives Ben more grain. Finally, the xconcept asks Leo if he can kick in more kernels for experiments.

So, the above xconcept performs conceptual processing that creates a new xconcept that will implement and test Ben's thought that he might create bubbles in bread by adding beer foam to dough. As a result of processing the above xconcept, the system displays:

1...31 Ben asks can you kick in more kernels for experiments?.

Leo needs to disambiguate Ben's question in four ways: 1) disambiguate "you" as a reference to himself; 2) disambiguate "kernels" as "grain"; 3) disambiguate "kick in" as a common English metaphor meaning "give to"; 4) disambiguate the unspecified object of "to" as Ben. Ben's metaphorical wording of his request is built into his xconcept, so that Leo will need to perform disambiguation, again since one goal of the simulation is to illustrate disambiguation.

Leo's first disambiguates "you" by applying the construction described above in step 1...4. The system displays this as:

> **1...31 (Leo translates as) Ben asks can Leo kick in more kernels for experiments?.**

Next, Leo disambiguates "kernel", using an xconcept with the following logic:

```
If A asks can I kick in more X for experiments
Then
   If X can mean Y
      And G is in current-domains
      And G can mean Y
      And I want A to make G edible
      And I have excess G
   Then
         think A asks can I kick in more G for experiments
```

In processing this xconcept, A is bound to Ben and X is bound to "kernels". Leo finds that "kernels" can mean "seeds", and therefore binds Y to "seeds". Leo finds that "grain" satisfies all the conditions for G, binding G to "grain". As a result, Leo generates a mental speech act that the system displays as:

> **1...31 (mu) Leo thinks Ben asks can Leo kick in more grain for experiments?.**

Next, Leo applies a construction to disambiguate "kick in". The Tala mentalese for this construction is:

```
(subformtrans
      (wusage verb)
      (subj
            (kick
                  (wusage verb)
                  (subj (?s (wusage noun)))
                  (in  (?x (wusage noun)))
                  ))
      (obj
            (give
                  (wusage verb)
                  (subj (?s (wusage noun)))
                  (obj (?x (wusage noun)))
                  (to me)
                  ]
```

The default object "me" is a simplification, not adequate in general. For example, someone asking for a donation to a charity might say "Could you kick in twenty dollars?" meaning "give to the charity" rather than "give to me". This simplification may be considered a logical bridge supporting the demonstration, in place of additional commonsense knowledge and reasoning. It is removable in principle, but doing so would require additional design and coding, and is left as a topic for future research.

Applying this construction yields an internal mental event the system displays as:

> **1...31** **(Leo translates as) (mu) Leo thinks Ben asks**
> **can Leo give more grain to me for experiments?.**

Finally, Leo disambiguates the phrase "to me" as "to Ben", using the construct described above for disambiguating "me" in step 1...5. This yields:

> **1...31** **(Leo translates as) (mu) Leo thinks Ben asks**
> **can Leo give more grain to Ben for experiments?.**

Leo has an xconcept that may be described as:

```
If I think A asks can I give more X to A for experiments
Then
    If I want A to make X edible
        And X is current-domains
        And I have excess X
    Then
        Give more X to A
```

In processing this xconcept, A is bound to Ben and X is bound to "grain". Leo finds that he wants Ben to make grain edible, that grain is in current-domains for discussion, and that he has excess grain. As a result, Leo generates an effept (physical action) that the system displays as:

> **1...32** **Leo gives more grain to Ben.**

This effept resets the grain behavior system current state to its initial state, i.e. unprocessed grain that resembles nuts.

Now Ben performs the new xconcept that he created at the end of timestep 1...26 above, to test what happens if he performs all the steps to make flat bread, and mixes beer foam into the dough before it is baked. The system displays:

> **1...33** **Ben pounds grain.**
> **1...33** **Ben soaks grain in water.**
> **1...33** **Ben mashs grain.**
> **1...33** **Ben mixs the dough with beer foam.**

1...33 Ben bakes dough.

1...33 Ben says Leo try this leavened bread.

When these effepts are processed, the grain behavior system responds as before to Ben's effepts for pounding, soaking, and mashing grain. The new effept for mixing beer foam into dough causes the dough to change state to become "leavened dough". When leavened dough is baked, it changes state to become "leavened bread".

In response to Ben's request to try the leavened bread, Leo generates an effept the system displays as:

1...34 Leo tries to eat bread.

The leavened bread (i.e. the Tala behavior system for grain) receives Leo's effept to eat bread as an input-action, and processes this with a finite-state behavior rule, which generates an output-action that Leo receives as a percept, in a Tala mentalese expression saying that bread is edible, thick, soft, tastes good, and not gooey. Leo again uses his general xconcept that if he perceives something, he will say what he perceives:

1...36 Leo says bread is edible, thick, soft, tastes good, and
not gooey.

To conclude the demonstration, Ben has a logical bridge xconcept with the description:

```
If someone says bread is edible, thick, soft, tastes
    good, and not gooey.
Then say Eureka!
```

And so, the final step of this instance of the discovery of bread demonstration displays:

1...37 Ben says Eureka!

Bibliography

Titles of books and papers are in italics, while names of proceedings and journals are not italicized.

Albus, J. S. (2011). *Path to a Better World: A Plan for Prosperity, Opportunity, and Economic Justice.* iUniverse, Inc.

Albus, J. S. & A. M. Meystel (2001). *Engineering of Mind: An Introduction to the Science of Intelligent Systems.* Wiley-Interscience.

Alderson-Day, B., S. Weis, S. McCarthy-Jones, P. Moseley, D. Smailes, & C. Fernyhough (2016). *The brain's conversation with itself: neural substrates of dialogic inner speech.* Social Cognitive and Affective Neuroscience, 2016, 110-120.

Aleksander, I. (1992). *Capturing consciousness in neural systems.* Artificial Neural Networks, 2. Proc. ICANN-92, North-Holland, 17-22.

———— (1996). *Impossible Minds: My Neurons, My Consciousness.* Imperial College Press, London.

———— (2001). *How to Build a Mind: Toward Machines with Imagination.* Columbia University Press.

———— (2005). *The World In My Mind, My Mind In The World.* Imprint Academic.

Aleksander, I. & B. Dunmall (2003). *Axioms and Tests for the Presence of Minimal Consciousness in Agents.* Journal of Consciousness Studies, 10, 4-5, 7-18.

Aleksander, I. & H. Morton (2007). *Depictive Architectures for Synthetic Phenomenology.* In *Artificial Consciousness,* 67-81, ed. A. Chella & Riccardo Manzotti (2007), Imprint Academic.

Anderson, J. R. & C. Lebiere (1998). *The Atomic Components of Thought.* Psychology Press.

Anderson, K. R., T. J. Hickey, & P. Norvig (2006). *JScheme.* 7.2 Release.

310

Anderson, S. L. (2005) *Asimov's 'Three Laws of Robotics' and machine metaethics*. In *Machine Ethics: Papers from the AAAI Fall Symposium*, ed. Anderson, M., Anderson, S., & Armen, C. (2005). Technical Report FS-05-06. AAAI Press.

Asimov, I. (1950). *I, Robot*. Bantam Books.

——— (1976). *The Bicentennial Man and Other Stories*. Doubleday.

Atkin, A. (2010). *Peirce's Theory of Signs*. In *The Stanford Encyclopedia of Philosophy* (Winter 2010 Edition), ed. E. N. Zalta.

Austin, J. L. (1962). *How To Do Things With Words: The William James Lectures delivered at Harvard University in 1955*. 2nd Edition, ed. J. O. Urmson & M. Sbisà. Harvard University Press.

Baars, B. J. (1995). *Can Physics Provide a Theory of Consciousness? A Review of Shadows of the Mind by Roger Penrose*. Psyche, 2, 8.

——— (1996). *Understanding Subjectivity: Global Workspace Theory and the Resurrection of the Observing Self*. Journal of Consciousness Studies, 3, 3, 211-216.

Baars, B. J. & N. M. Gage (2007). *Cognition, Brain, and Consciousness: Introduction to Cognitive Neuroscience*. Elsevier.

Bach, J. (2009). *Principles of Synthetic Intelligence–PSI: An Architecture of Motivated Cognition*. Oxford University Press.

——— (2015). *Modeling motivation in MicroPsi 2*. Artificial General Intelligence, 8th International Conference, AGI 2015, 3-13.

Bachwerk, M. & C. Vogel (2011). *Establishing Linguistic Conventions in Task-Oriented Primeval Dialog*. In *Analysis of Verbal and Nonverbal Communication and Enactment: The Processing Issues, Lecture Notes in Computer Science*, ed. A. Esposito, A. Vinciarelli, K. Vicsi, C. Pelauchaud, & A. Nijholt, Vol. 6800, 48-55. Springer.

Bar-Haim, R., I. Dagan, W. Dolan, L. Ferro, D. Giampiccolo, B. Magnini, & I. Szpektor (2006). *The Second PASCAL Recognising Textual Entailment Challenge*. Proceedings of the Second PASCAL Challenges Workshop on Recognising Textual Entailment, pp.1-9.

Bibliography

Barsalou, L. W. (1993). *Flexibility, Structure, and Linguistic Vagary in Concepts: Manifestations of a Compositional System of Perceptual Symbols*. In *Theories of Memory*, ed. A. F. Collins, S. E. Gathercole, M. A. Conway, & P. E. Morris, 29-101. Lawrence Erlbaum.

―――― (1999). *Perceptual symbol systems*. Behavioral and Brain Sciences, 22, 577-660.

―――― (2012). *The Human Conceptual System*. In *The Cambridge Handbook of Psycholinguistics*, ed. M. J. Spivey, K. McRae, & M. F. Joanisse, 239-258. Cambridge University Press.

Barwise, J. & R. Cooper (1981). *Generalized quantifiers and natural language*. Linguistics and Philosophy 4: 159-219.

Bateman, J. A., R. Henschel, & F. Rinaldi (1995). *The Generalized Upper Model 2.0*. German National Research Center for Information Technology, Darmstadt.

Bello, P. & S. Bringsjord (2013) *On how to build a moral machine*. Topoi, 32, 2, 1-25.

Benedetti, G. (2011). *An Enigma in Languag: The Meaning of the Fundamental Linguistic Elements. A Possible Explanation in Terms of Cognitive Functions: Operational Semantics*. Nova Science Publishers.

Bergen, B. K., N. C. Chang, & S. Narayan (2004*). Simulated action in an Embodied Construction Grammar*. Proceedings Twenty-Fifth Annual Conference of the Cognitive Science Society.

Berwick, R. C. & N. Chomsky (2016). *Why Only Us: Language and Evolution*. MIT Press.

Biber, D., S. Johansson, G. Leech, S. Conrad, & E. Finegan (1999). *Longman Grammar of Spoken and Written English*. Longman Publishers, London.

Blackmore, S. (2011). *Consciousness: An Introduction*. Oxford University Press.

Boden, M. A. (1983). *Artificial intelligence as a humanizing force*. Proceedings International Joint Conference on Artificial Intelligence 1983, 1197-1198.

———— (2004). *The Creative Mind: Myths and Mechanisms*. Routledge.

Bohm, C. & G. Jacopini (1966). *Flow diagrams, Turing machines and languages with only two formation rules*. Communications ACM, 9, 5, 366-371.

Boroditsky, L. & J. Prinz (2008). *What thoughts are made of*. In *Embodied Grounding: Social, Cognitive, Affective, and Neuroscientific Approaches*, ed. G. Semin and E. Smith, 108-125. Cambridge University Press.

Bos, J. & K. Markert (2006). *When logical inference helps determining textual entailment (and when it doesn't)*. Proceedings of the Second PASCAL Challenges Workshop on Recognizing Textual Entailment.

Bosse, T. & J. Treur (2006). *Formal Interpretation of a Multi-Agent Society As a Single Agent*. Journal of Artificial Societies and Social Simulation, 9, 2.

Bostock, J. & H. T. Riley (1856). *The Natural History of Pliny*. London: H. G. Bohn.

Bostrom, N. (2014). *Superintelligence: Paths, Dangers, Strategies*. Oxford University Press.

Brain, M. (2013). *Robotic Nation and Robotic Freedom*. Tenth Anniversary Edition, BYG Publishing.

Bringsjord, S., K. Arkoudas, & P. Bello (2006). *Toward a general logicist methodology for engineering ethically correct robots*. IEEE Intelligent Systems, July 2006, 38-44.

Brooks, R. A. (1997). *From Earwigs to Humans*. Robotics and Autonomous Systems, 20, 2-4, 291-304.

Brynjolfsson, E. & A. McAfee (2011). *Race Against The Machine: How the Digital Revolution is Accelerating Innovation, Driving Productivity, and Irreversibly Transforming Employment and the Economy*. Digital Frontier Press.

Buescu, J., D. S. Graça, & N. Zhong (2011). *Computability and dynamical systems*. In *Dynamics, Games and Science*, ed. M. M. Peixoto, A. A. Pinto, D. A. J. Rand, 169-181. Springer.

Buitelaar, P., B. Magnini, C. Strapparava, & P. Vossen (2006). *Domain-specific WSD*. In *Word Sense Disambiguation: Algorithms and Applications*, ed. E. Agirre and P. Edmonds, 275-298. Springer.

Bunt, H. C. (1994). *Context and dialogue control*. THINK Quarterly, 3, 1, 19-31.

———— (2000). *Dialogue Pragmatics and Context Specification*. In *Abduction, Belief and Context in Dialogue: Studies in Computational Pragmatics*, ed. H. C. Bunt & W. J. Black, 81-150. John Benjamins Publishing Co.

———— (2007). *The Semantics of Semantic Annotation*. Proceedings 21st Pacific Asia Conference on Language, Information and Computation, Seoul, Korea, November 1-3, 2007.

———— (2008). *Semantic Underspecification: Which Technique for What Purpose?* In *Computing Meaning*, ed. H. Bunt and R. Muskens, Volume 3, 55-85. Springer Science + Business Media B.V.

———— (2009). *Semantic Annotations as Complementary to Underspecified Semantic Representations*. Proceedings 8th International Conference on Computational Semantics (IWCS-8), Tilburg, January 7-9, 2009, 33-45.

Bunt, H. C. & W. J. Black (2000). *The ABC of Computational Pragmatics*. In *Abduction, Belief and Context in Dialogue: Studies in Computational Pragmatics*, ed. H. C. Bunt & W. J. Black, 1-46. John Benjamins Publishing Co.

Bunt, H. C. & W. J. Black, eds. (2000). *Abduction, Belief and Context in Dialogue: Studies in Computational Pragmatics*. John Benjamins Publishing Co.

Burgener, R. (2006). *20Q: the Neural Network Mind Reader*. Engineering Colloquium announcement, Goddard Space Flight Center.

Carter, R. & M. McCarthy (2006). *Cambridge Grammar of English: A Comprehensive Guide. Spoken and Written English Grammar and Usage*. Cambridge University Press.

Cassimatis, N. L. (2002). *Polyscheme: A Cognitive Architecture for Integrating Multiple Representation and Inference Schemes*. Doctoral Dissertation, Media Laboratory, Massachusetts Institute of Technology.

———— (2006). *A Cognitive Substrate for Human-Level Intelligence*. AI Magazine, 27, 2, 45-56.

Chalmers, D. J. (1995a). *Facing Up to the Problem of Consciousness*. Journal of Consciousness Studies, 2, 3, 200-219.

———— (1995b). *Minds, Machines, and Mathematics*. Psyche, 2, 11-20.

———— (1995c). *Absent Qualia, Fading Qualia, Dancing Qualia*. In *Conscious Experience*, ed. T. Metzinger. Imprint Academic.

———— (1996). *The Conscious Mind: In Search of a Fundamental Theory*. Oxford University Press.

———— (2010). *The Character of Consciousness*. Oxford University Press.

———— (2011). *A computational foundation for the study of cognition*. Journal of Cognitive Science, 12, 4, 323-357.

Chang, N. C. (2008). *Constructing Grammar: A Computational Model of the Emergence of Early Constructions*. Ph.D. Thesis, University of California at Berkeley.

Charlesworth, A. (2016). *A theorem about computationalism and "absolute" truth*. Minds & Machines, 26, 205-226.

Chella, A., M. Frixione, & S. Gaglio (1997). *A Cognitive Architecture for Artificial Vision*. Artificial Intelligence, 89, 73-111.

———— (1998). *An Architecture for Autonomous Agents Exploiting Conceptual Representations*. Robotics and Autonomous Systems, 25, 231-240.

Chomsky, N. (1965). *Aspects of the Theory of Syntax*. MIT Press.

———— (1966). *Cartesian Linguistics: A Chapter in the History of Rationalist Thought*. Harper & Row.

———— (1972). *Language and Mind*. Harcourt Brace Jovanovich.

———— (1975). *Reflections on Language*. Pantheon.

———— (1995). *The Minimalist Program*. MIT Press.

———— (2000). *New Horizons in the Study of Language and Mind*. Cambridge University Press.

———— (2015). *The Minimalist Program*. 20th Anniversary Edition. MIT Press.

Close, F. (2009). *Nothing: A Very Short Introduction*. Oxford University Press.

Cole, D. (2009). *The Chinese Room argument*. In *The Stanford Encyclopedia of Philosophy* (Winter 2009 Edition), ed. E. N. Zalta.

Colyvan, M. (2011). *Indispensability Arguments in the Philosophy of Mathematics*. In *The Stanford Encyclopedia of Philosophy* (Spring 2011 Edition), ed. E. N. Zalta.

Cooper, R. *et al.* (1996). *Using the framework*. Technical Report LRE 62-051 D-16. The FraCaS Consortium. University of Edinburgh, School of Informatics.

Copeland, B. J. (2002). *Hypercomputation*. Minds and Machines, 12, 461-502.

Coulardeau, J., & I. Eve (2017) *Cro-Magnon's Language*. Editions La Dondaine

Coven, H. J. (1991). *A Descriptive-Operational Semantics for Prescribing Programming Languages with "Reflective" Capabilities*. Ph.D. Thesis, Arizona State University.

Craik, K. J. W. (1943). *The Nature of Explanation*. Cambridge University Press.

Croft, W. (2002). *Radical Construction Grammar: Syntactic Theory in Typological Perspective*. Oxford University Press.

Croft, W., & D. A. Cruse (2004). *Cognitive Linguistics*. Cambridge University Press.

Cutrer, L. M. (1994). *Time and Tense in Narrative and in Everyday Life*. Doctoral thesis, Department of Cognitive Science, University of California, San Diego.

Daelemans, W., K. De Smedt, & G. Gazdar (1992). *Inheritance in Natural Language Processing*. Computational Linguistics, 18, 2, 205-218.

Daelemans, W. & K. De Smedt (1994). *Default Inheritance in an Object-Oriented Representation of Linguistic Categories*. International Journal of Human-Computer Studies 41, 149-177.

Davis, M. (1990). *Is Mathematical Insight Algorithmic?* Behavioral and Brain Sciences, 13, 4, 659-660.

—————— (1993). *How Subtle is Gödel's Theorem? More on Roger Penrose.* Behavioral and Brain Sciences, 16, 3, 611-612.

—————— (2006). *The Myth of Hypercomputation.* In *Alan Turing: Life and Legacy of a Great Thinker*, ed. Christof Teuscher. Springer.

—————— (2006). *Why there is no such discipline as hypercomputation.* Applied Mathematics and Computation, 178, 4-7.

Dennett, D. C. (1987). *The Intentional Stance.* The MIT Press.

—————— (1991). *Consciousness Explained.* Little, Brown & Co.

—————— (1996). *Kinds of Minds: Toward an Understanding of Consciousness.* Basic Books.

—————— (2005). *Sweet Dreams: Philosophical Obstacles to a Science of Consciousness.* The MIT Press.

Dolan, W. B., L. Vanderwende, & S. D. Richardson (1993). *Automatically deriving structured knowledge bases from online dictionaries.* Proceedings Pacific Assoc. for Computational Linguistics, 1993; Tech. Report MSR-TR-93-06, Microsoft, Redmond, Washington.

Doyle, J. (1980). *A Model for Deliberation, Action, and Introspection.* Ph.D. Thesis. MIT AI Lab TR-581.

—————— (1983). *A Society of Mind: Multiple perspectives, reasoned assumptions, and virtual copies.* Proceedings International Joint Conference on Artificial Intelligence 1983, 309-314.

Dretske, F. (1985). *Machines and the Mental.* Presidential Address to the 83rd Annual Meeting of the Western Division of the American Philosophical Association. APA Proceedings, 59, 23-33.

Dreyfus, H. L. (1992). *What Computers Still Can't Do: A Critique of Artificial Reason.* The MIT Press.

Duchin, F. (1983). *Computers, employment, and the distribution of income.* Proceedings International Joint Conference on Artificial Intelligence 1983, 1196-1197.

———— (1998). *Structural Economics: Measuring Change in Technology, Lifestyles, and the Environment.* Island Press, Washington, D.C.

Easterly, W. (2001). *The Elusive Quest for Growth: Economists' Adventures and Misadventures in the Tropics.* MIT Press.

Eichengreen, B. (2011). *A critique of pure gold.* The National Interest, August 21, 2011.

Evans, R., A. Gelbukhy, G. Grefenstettez, P. Hanks, M. Jakubícek, D. McCarthy, M. Palmer, T. Pedersen, M. Rundell, P. Rychlý, S. Sharoff, & D. Tugwell (2016). *Adam Kilgarriff's legacy to computational linguistics and beyond.* CICLing 2016: Computational Linguistics and Intelligent Text Processing, 3-25. Springer

Evans, V. (2009). *How Words Mean: Lexical Concepts, Cognitive Models and Meaning Construction.* Oxford University Press.

Evans, V. & M. Green (2006). *Cognitive Linguistics: An Introduction.* Lawrence Erlbaum Associates.

Fahlman, S. E. (1979). *NETL: A System for Representing and Using Real World Knowledge.* MIT Press.

Fauconnier, G. (1984). *Espaces mentaux.* Editions de Minuit, Paris.

———— (1985). *Mental Spaces: Aspects of Meaning Construction in Natural Language.* MIT Press.

———— (1994). *Mental Spaces: Aspects of Meaning Construction in Natural Language.* Augmented paperback edition. Cambridge University Press.

———— (1997). *Mappings in Thought and Language.* Cambridge University Press.

Fauconnier, G. & E. Sweetser, eds. (1996). *Spaces, Worlds, and Grammars.* University of Chicago Press.

Fauconnier, G. & M. Turner (1994). *Conceptual Projection and Middle Spaces.* Technical Report 9401, Department of Cognitive Science, University of California, San Diego.

———— (1998). *Conceptual integration networks.* Cognitive Science, 22, 133-187.

———— (2002). *The Way We Think: Conceptual Blending and the Mind's Hidden Complexities.* Basic Books, New York.

Feferman, S. (1995). *Penrose's Gödelian argument: A Review of Shadows of the Mind.* PSYCHE: An Interdisciplinary Journal of Research On Consciousness, 2, 21-32.

———— (2006). *Are There Absolutely Unsolvable Problems? Gödel's Dichotomy.* Philosophia Mathematica, 14, 2, 134-152.

———— (2011). *Gödel's incompleteness theorems, free will and mathematical thought.* In *Free Will and Modern Science*, ed. R. Swinburne, 102-122. Oxford University Press, for the British Academy.

Feldman, J. A. (2002). *A Proposed Formalism for ECG Schemas, Constructions, Mental Spaces, and Maps.* International Institute of Computer Science, Tech. Rep. TR-02-010, Sep. 2002.

———— (2006). *From Molecule to Metaphor: A Neural Theory of Language.* MIT Press.

Fernyhough, C. (2016). *The Voices Within: The History and Science of How We Talk to Ourselves.* Basic Books.

———— (2017). *Talking to ourselves.* Scientific American, August 2017, pp.76–79.

Ferrucci, D., E. Brown, J. Chu-Carroll, J. Fan, D. Gondek, A. A. Kalyanpur, A. Lally, J. W. Murdock, E. Nyberg, J. Prager, N. Schlaefer, & C. Welty (2010). *Building Watson: An Overview of the DeepQA Project.* AI Magazine, 31, 3, 59-79.

Fillmore, C. J. (1975). *An alternative to checklist theories of meaning.* Proceedings First Annual Meeting of the Berkeley Linguistics Society. Amsterdam: North Holland, 123-31.

———— (1977). *Scenes-and-frame semantics.* In *Linguistic Structures Processing*, ed. A. Zampolli, 55-82. Amsterdam: North Holland.

———— (1978). *On the Organization of Semantic Information in the Lexicon.* In C. J. Fillmore (2003) *Form and Meaning in Language, Volume I: Papers on Semantic Roles.* CSLI Publications, Stanford University.

———— (1985). *Frames and the semantics of understanding.* Quaderni di Semantica, 6, 222-54.

———— (1988) *The mechanisms of construction grammar.* Proceedings Berkeley Linguistics Society, 14, 35-55.

———— (2002). *Lexical Isolates.* In *Lexicography and Natural Language Processing: A Festscrift in Honour of B.T.S. Atkins,* ed. M-H. Corréard, 105-124. Euralex 2002.

Fillmore, C. J., P. Kay, & M. K. O'Connor (1988). *Regularity and idiomaticity: the case of let alone.* Language, 64, 3, 501-38.

Fillmore, C. J. & B. T. Atkins (1992). *Toward a frame-based lexicon: the semantics of RISK and its neighbors.* In *Frames, Fields and Contrasts,* ed. A. Lehrer & E. F. Kittay, 75-102. Lawrence Erlbaum.

Fleck, A. (2001). *Formal Models of Computation: The Ultimate Limits of Computing.* AMAST Series in Computing, Vol. 7. World Scientific Publishing.

Flickinger, D., S. Oepen, & G. Ytrestøl (2010). *WikiWoods: Syntacto-Semantic Annotation for English Wikipedia.* Proceedings International Conference on Language Resources and Evaluation, LREC 2010, 1665-1671.

Fodor, J. A. (1975). *The Language of Thought.* T. Y. Crowell Co., New York.

———— (2008). *LOT 2: The Language of Thought Revisited.* Oxford University Press.

Forbus, K. D. & T. R. Hinrichs (2006). *Companion Cognitive Systems: A Step toward Human-Level AI.* AI Magazine. Volume 27 Number 2, 83-95.

Ford, M. (2009). *The Lights in the Tunnel: Automation, Accelerating Technology and the Economy of the Future.* Acculant Publishing.

———— (2013). *Could Artificial Intelligence Create an Unemployment Crisis?* Communications ACM, 56, 7, 37-39.

Franzén, T. (2005). *Gödel's Theorem: An Incomplete Guide to Its Use and Abuse*. A. K. Peters, Ltd.

Friedman, M. (1962). *Capitalism and Freedom*. University of Chicago Press.

Gamow, G. & M. Stern (1958). *Puzzle-Math*. Viking Press.

Gärdenfors, P. (1995). *Three levels of inductive inference*. Studies in Logic and the Foundations of Mathematics, 134, 427-449. Elsevier.

———— (2000). *Conceptual Spaces: The Geometry of Thought*. MIT Press.

Gates, E. (2009.) *Einstein's Telescope: The Hunt for Dark Matter and Dark Energy in the Universe*. W. W. Norton & Company.

Gelernter, D. (1994). *The Muse in the Machine: Computerizing the Poetry of Human Thought*. The Free Press, Simon & Schuster.

Giampiccolo, D., B. Magnini, I. Dagan, & W. Dolan (2007). *The Third PASCAL Recognizing Textual Entailment Challenge*. Proceedings of the ACL-07 Workshop on Textual Entailment and Paraphrasing.

Gliozzo, A. M., B. Magnini, & C. Strapparava (2004). *Unsupervised domain relevance estimdtion for word sense disambiguation*. In Proceedings 2004 Conference on Empirical Methods in Natural Language Processing. 380-387.

Gliozzo, A. M. & C. Strapparava (2009). *Semantic Domains in Computational Linguistics*. Springer-Verlag.

Goddard, C. & A. Wierzbicka, eds. (2002). *Semantic and Lexical Universals: Theory and Empirical Findings*. John Benjamins, Amsterdam.

Gödel, K. (1951.) *Some basic theorems on the foundations of mathematics and their implications*. In *Collected works / Kurt Gödel*, III. ed. S. Feferman, et al., 1995, 304-23. Oxford University Press.

Goertzel, B. (2011). *Self-Programming = Learning about Intelligence-Critical System Features*. AGI-11 Workshop on Self-Programming in AGI Systems.

Goertzel, B. & C. Pennachin, eds. (2007). *Artificial General Intelligence*. Springer.

Goff, P. (2017). *Consciousness and Fundamental Reality.* Oxford University Press.

———— (2018). *Is the Universe a conscious mind?* Aeon, February 8, 2018.

https://aeon.co/essays/cosmopsychism-explains-why-the-universe-is-fine-tuned-for-life

Goldberg, A. E. (1995). *Constructions: A Construction Grammar Approach to Argument Structure.* Chicago University Press.

———— (2006). *Constructions at Work: The Nature of Generalization in Language.* Oxford University Press.

Good, I. J. (1965). *Speculations concerning the first ultraintelligent machine.* Advances in Computers, vol. 6.

Gosling, J., W. Joy, G. Steele, & G. Bracha (2005). *The Java™ Language Specification.* Third Edition. Addison-Wesley.

Grady, J., T. Oakley, & S. Coulson (1999). *Blending and metaphor.* In *Metaphor in Cognitive Linguistics,* ed. R.W. Gibbs and G. Steen, 101-124. Amsterdam: John Benjamins.

Grice, H. P. (1989). *Studies in the Way of Words.* Harvard University Press.

Grosz, B. J. & P. Stone (2018). *A century-long commitment to assessing Artificial Intelligence and its impact on society.* Communications ACM, 61, 12, 68-73.

Gubrud, M. A. (1997). *Nanotechnology and international security.* Fifth Foresight Conference on Molecular Nanotechnology.

Hall, J. S. (2011). *Self-Programming through Imitation.* AGI-11 Workshop on Self-Programming in AGI Systems.

Hameroff, S. (1998). *Quantum computation in brain microtubules? The Penrose-Hameroff "Orch OR" model of consciousness.* Philosophical Transactions of the Royal Society (London) Series A, 356, 1869-1896.

Hameroff, S. & R. Penrose (2014). *Consciousness in the universe: A review of the 'Orch OR' theory.* Physics of Life Reviews, 11, 1, 39-78.

Harari, Y. N. (2015). *Sapiens: A Brief History of Humankind*. Harper-Collins Publishers.

Harel, D. (1980). *On Folk Theorems*. Communications ACM, 23, 7, 379-389.

Hart, N. (1979). *SIMON: Syntactically Intelligent Memory Oriented Networks*, Senior Thesis, Information Sciences, University of California at Santa Cruz. Unpublished.

Haugeland, J. (1985). *Artificial Intelligence: The Very Idea*. MIT Press.

Hawkins, J. & S. Blakeslee (2004). *On Intelligence*. Times Books.

Hinton, G. E. (2006). *Learning multiple layers of representation*. Trends in Cognitive Sciences, 11, 10, 428-434.

Hobbs, J. R. (1983). *An Improper Treatment of Quantification in Ordinary English*. Proceedings of the 21st Annual Meeting, Association for Computational Linguistics, 57-63.

———— (2003). *The Logical Notation: Ontological Promiscuity*. Manuscript. http://www.isi.edu/~hobbs/disinf-chap2.pdf

———— (2004). *Abduction in Natural Language Understanding*. In *Handbook of Pragmatics*, ed. L. Horn and G. Ward, 724-741, Blackwell.

Hobbs, J. R., M. Stickel, D. Appelt, & P. Martin (1993). *Interpretation as Abduction*. Artificial Intelligence, 63 (1-2): 69-142.

Hodges, A. (2011). *Alan Turing*. In *The Stanford Encyclopedia of Philosophy* (Summer 2011 Edition), ed. E. N. Zalta.

Hofstadter, D. R. (1995). *Fluid Concepts and Creative Analogies: Computer Models of the Fundamental Mechanisms of Thought*. BasicBooks, New York.

Holland, J. H., K. J. Holyoak, R. E. Nisbett, & P. R. Thagard (1986). *Induction: Processes of Inference, Learning, and Discovery*. MIT Press.

Horn, L. R. (2001). *A Natural History of Negation*. Second Edition. CSLI Publications, Center for the Study of Language and Information, Stanford University.

Huddleston, R. & G. K. Pullum (2002). *The Cambridge Grammar of the English Language*. Cambridge University Press.

—— (2005). *A Student's Introduction to English Grammar*. Cambridge University Press.

Hudson, R. A. (1984). *Word Grammar*. Blackwell.

—— (1990). *English Word Grammar*. Blackwell.

—— (1999). *Subject-verb agreement in English*. English Language and Linguistics, 3, 173-207.

—— (2007). *Language Networks: The New Word Grammar*. Oxford University Press.

—— (2010). *An Introduction to Word Grammar*. Cambridge University Press.

Jackendoff, R. (1983). *Semantics and Cognition*. MIT Press.

—— (1989). *What Is a Concept, That a Person May Grasp It?* Mind & Language, 4, 1-2, 68-102, Wiley-Blackwell. Also in Jackendoff, R. (1992) *Languages of the Mind*, MIT Press.

—— (2007). *Language, Consciousness, Culture:Essays on Mental Structure*. MIT Press.

Jackson, P. C. (1974). *Introduction to Artificial Intelligence*. Petrocelli/Charter Publishers.

—— (1979). *Concept: A Context for High-Level Descriptions of Systems Which Develop Concepts*. Master's Thesis, University of California at Santa Cruz.

—— (1985). *Introduction to Artificial Intelligence*. Second Edition. Dover Publications.

—— (1992). *Proving unsatisfiability for problems with constant cubic sparsity*. Artificial Intelligence, 57, 125-137.

—— (2014). *Toward Human-Level Artificial Intelligence: Representation and Computation of Meaning in Natural Language*. Doctoral Thesis,

Tilburg University, The Netherlands. Tilburg Center for Cognition and Communication (TiCC) Ph.D. Series No. 32. Dutch Research School for Information and Knowledge Systems (SIKS) Dissertation Series No. 2014-09.

——— (2017). *Toward human-level models of minds*, 2017 AAAI Fall Symposium Series Technical Reports, FS-17-05, 371-375.

——— (2018a). *Toward beneficial human-level AI... and beyond*, 2018 AAAI Spring Symposium Series Technical Reports, SS-18-01, 48-53.

——— (2018b). *Postscript for 'beneficial human-level AI... and beyond.'* TalaMind White Paper, TalaMind LLC.

——— (2018c). *The intelligence level and TalaMind.* Sixth Annual Conference on Advances in Cognitive Systems, Poster Session.

——— (2018d). *Natural language in the Common Model of Cognition.* Procedia Computer Science, 145, 699-709.

——— (2018e). *Thoughts on bands of action.* Procedia Computer Science, 145, 710-716.

——— (2019). *Introduction to Artificial Intelligence.* Third Edition. Dover Publications, New York.

James, W. (1890). *The Principles of Psychology.* Henry Holt & Co.

Johnson, M. (1987). *The Body in the Mind: The Bodily Basis of Meaning, Imagination and Reason.* University of Chicago Press.

Johnson-Laird, P. N. (1983). *Mental Models: Towards a Cognitive Science of Language, Inference, and Consciousness.* Harvard University Press.

——— (2006) *How We Reason.* Oxford University Press.

——— (2010). *Mental models and human reasoning.* Proceedings of the National Academy of Sciences, 107, pp.18243-18250.

——— (2013). *The mental models perspective.* In *The Oxford Handbook of Cognitive Psychology*, ed. D. Reisberg. Oxford University Press.

Joy, W. (2000). *Why the future doesn't need us.* Wired Magazine, Issue 8.04, April 2000.

Kay, C. & M. L. Samuels (1975). *Componential analysis in semantics: its validity and applications.* Transactions Philological Society, 49-81.

Kay, P. & C. Fillmore (1999). *Grammatical constructions and linguistic generalizations: the What's X doing Y construction,* Language, 75, 1-34.

Keynes, J. M. (1930). *Economic Possibilities for our Grandchildren.* In *Essays in Persuasion,* Volume 9 of *The Collected Writings of John Maynard Keynes,* ed. E. Johnson & D. Moggridge, 2013. Cambridge University Press.

Kilgarriff, A. (1997). *"I don't believe in word senses."* Computers and the Humanities, 31: 91-113.

——— (2007). *Word senses. In Word Sense Disambiguation: Algorithms and Applications,* ed. E. Agirre & P. Edmonds, 29-46. Springer.

Knuth, D. E. (2002). *Selected Papers on Computer Languages.* Center for the Study of Language and Information, Stanford University.

Knuth, D. E. & R. W. Floyd (1971). *Notes on avoiding 'go to' statements.* Information Processing Letters, 1, 23-31. Errata, p.177. Reprinted as Chapter 23 in *Selected Papers on Computer Languages.*

Kotseruba, I. & J. K. Tsotsos (2018). *40 years of cognitive architecture research: Core cognitive abilities and practical applications.* Artificial Intelligence Review, 1-78. Springer.

Kövecses, Z. & G. Radden (1998). *Metonymy: developing a cognitive linguistic view.* Cognitive Linguistics, 9, 1, 37-77.

Kralik, J. D. (2017). *Architectural design of mind & brain from an evolutionary perspective.* 2017 AAAI Fall Symposium Series Technical Reports, FS-17-05, 394-400.

Kralik, J. D., J. H. Lee, P. S. Rosenbloom, P. C. Jackson, S. L. Epstein, O. J. Romero, R. Sanz, O. Larue, H. Schmidtke, S. W. Lee, & K. McGreggor (2018). *Metacognition for a Common Model of Cognition,* Procedia Computer Science, 145, 730-739.

Kuipers, B. (2018). *How can we trust a robot?* Communications of the ACM, 61, 3, 86-95.

Kurzweil, R. (2005). *The Singularity Is Near: When Humans Transcend Biology*. Viking.

Laird, J. E. (2008). *Extending the Soar Cognitive Architecture*. Proceedings First Conference on Artificial General Intelligence (AGI-08).

Laird, J. E., A. Newell, & P. S. Rosenbloom (1987). *Soar: An architecture for general intelligence*. Artificial Intelligence, 33: 1-64.

Laird, J. E., C. Lebiere, & P. S. Rosenbloom (2017). *A Standard Model of the Mind: Toward a common computational framework across artificial intelligence, cognitive science, neuroscience, and robotics*. AI Magazine, Winter 2017, 38, 4, 13-26..

Lakoff, G. (1970). *Linguistics and natural logic*. Synthese 22:151-271.

———— (1987). *Women, Fire, and Dangerous Things: What Categories Reveal about the Mind*. University of Chicago Press.

Lakoff, G. & M. Johnson (1980). *Metaphors We Live By*. University of Chicago Press.

———— (1999). *Philosophy in the Flesh: The Embodied Mind and Its Challenge to Western Thought*. Basic Books.

Lakoff, G. & M. Turner (1989). *More than Cool Reason: A Field Guide to Poetic Metaphor*. University of Chicago Press.

Langacker, R. W. (1987). *Foundations of Cognitive Grammar: Volume I: Theoretical Prerequisites*. Stanford University Press.

———— (1991). *Foundations of Cognitive Grammar: Volume 2: Descriptive Application*. Stanford University Press.

———— (2008). *Cognitive Grammar: A Basic Introduction*. Oxford University Press.

Langley, P. (2006). *Cognitive Architectures and General Intelligent Systems*. AI Magazine, 27, 2, 33-44.

Langley, P., D. Choi, & S. Rogers (2009). *Acquisition of hierarchical reactive skills in a unified cognitive architecture*. Cognitive Systems Research, 10, 316–332.

Bibliography

Langley, P., H. A. Simon, G. L. Bradshaw, & J. M. Zytkow (1987). *Scientific Discovery: Computational Explorations of the Creative Processes.* MIT Press.

Larue, O., R. West, P. S. Rosenbloom, C. L. Dancy, A. V. Samsonovich, D. Petters, & I. Juvina (2018). *Emotion in the Common Model of Cognition.* Procedia Computer Science, 145, 740-746.

Leijnen, S. (2011). *Thinking Outside the Box: Creativity in Self-Programming Systems.* AGI-11 Workshop on Self-Programming in AGI Systems.

Lenat, D. B. (1976). *AM: An Artificial Intelligence Approach to Discovery in Mathematics as Heuristic Search.* Ph.D. Thesis, Stanford University.

———— (1995). *Cyc: A large-scale investment in knowledge infrastructure.* Communications ACM, 38, 11, 33-38.

Lenat, D. B. & J. S. Brown (1984). *Why AM and EURISKO appear to work.* Artificial Intelligence 23(3):269-294.

Leontief, W. (1983a). *Technological Advance, Economic Growth, and the Distribution of Income.* Population and Development Review, 9, 3, 403-410.

———— (1983b). *National Perspective: The Definition of Problems and Opportunities.* In *The Long-Term Impact of Technology on Employment and Unemployment,* 3-7. National Academy of Engineering Symposium, June 30, 1983. National Academy Press, Washington, D. C.

Leontief, W. & F. Duchin (1986). *The Future Impact of Automation on Workers.* Oxford University Press.

Lindström, P. (2001). *Penrose's new argument.* Journal of Philosophical Logic, 30, 241-250.

———— (2006). *Remarks on Penrose's new argument.* Journal of Philosophical Logic, 35, 231-237.

Lucas, J. R. (1959). *Minds, Machines, and Gödel.* Paper presented to the Oxford Philosophical Society on October 30, 1959, and published in Philosophy, 36, 112-127, 1961.

——— (2003). *The Gödelian Argument: Turn Over the Page*. Etica & Politica / Ethics & Politics, 2003, 1.

——— (2011). *Feferman on Gödel and Free Will: a response to chapter 6*. In R. Swinburne, ed. (2011) *Free Will and Modern Science*. Oxford University Press / British Academy.

MacCartney, W. (2009). *Natural Language Inference*. Ph.D. Thesis, Stanford University.

MacCartney, W. & C. D. Manning (2008). *Modeling semantic containment and exclusion in natural language inference*. The 22nd International Conference on Computational Linguistics (Coling-08), Manchester, UK, August 2008.

——— (2009). *An extended model of natural logic*. The Eighth International Conference on Computational Semantics (IWCS-8), Tilburg, Netherlands, January 2009.

Mandler, J. M. (1988). *How to build a baby: On the development of an accessible representational system*. Cognitive Development, 3, 113-136.

——— (1992). *How to build a baby: II. Conceptual primitives*. Psychological Review, 99, 4, 587-604.

Markram, H. (2006). *The Blue Brain Project*. Nature Reviews Neuroscience, 7, February 2006, 153-160.

Masterman, M. (1961). *Semantic message detection for machine translation, using an interlingua*. First International Conference on Machine Translation of Languages and Applied Language Analysis, 438-475. National Physical Laboratory, Her Majesty's Stationery Office, London.

——— (2006). *Language, Cohesion and Form*. Edited by Y. Wilks. Cambridge University Press.

McCarthy, J. (1955) *Proposal for research by John McCarthy*. A section of McCarthy *et al.* (1955).

——— (1959). *Programs with common sense*. Proceedings of the Teddington Conference on the Mechanization of Thought Processes, Her Majesty's Stationary Office, 75-91.

—— (1960). *Recursive functions of symbolic expressions and their computation by machine, Part I.* Communications ACM, April 1960.

—— (1980). *Circumscription: a form of nonmonotonic reasoning.* Artificial Intelligence, 13, 27-39.

—— (1992). *Elephant 2000: a programming language based on speech acts.* Unpublished paper. Abstract published for Keynote Address in the 2007 Companion to the 22nd ACM SIGPLAN Conference on Object-Oriented Programming Systems and Applications, 723-724 .

—— (1995). *Awareness and understanding in computer programs: A review of Shadows of the Mind by Roger Penrose.* Psyche, 2, 11.

—— (2002). *Making robots conscious of their mental states.* Unpublished paper.

—— (2004). *The robot and the baby.* Unpublished short story.

—— (2005). *Beyond Lisp.* Unpublished lecture slides.

—— (2006). *Human-level AI is harder than it seemed in 1955.* Unpublished lecture slides.

—— (2007). *From here to human-level AI.* Artificial Intelligence, 171, 1174-1182.

—— (2008). *The well-designed child.* Artificial Intelligence, 172, 18, 2003-2014.

McCarthy, J., M. L. Minsky, N. Rochester, & C. E. Shannon (1955). *A Proposal for the Dartmouth Summer Research Project on Artificial Intelligence.* The first 5 pages were reprinted in AI Magazine, 2006, 27, 4, 12-14. The proposal was also reprinted in *Artificial Intelligence: Critical Concepts in Cognitive Science,* ed. R. Chrisley & S. Begeer (2000), 2, 44-53. Routledge Publishing.

http://jmc.stanford.edu/articles/dartmouth/dartmouth.pdf

McCarthy, J., P. W. Abrahams, D. J. Edwards, T. P. Hart, & M. I. Levin (1962). *LISP 1.5 Programmer's Manual.* Second Edition. MIT Press.

McCullough, D. (1995). *Can Humans Escape Gödel? A Review of Shadows of the Mind by Roger Penrose.* Psyche, 2, 4, 57-65.

McDermott, D. (1995). *[STAR] Penrose is wrong*. Psyche, 2, 1.

McDuff, D. & M. Czerwinski (2018). *Designing emotionally sentient agents*. Communications ACM, 61, 12, 74-83.

Mc Kevitt, P. & C. Guo (1996). *From Chinese Rooms to Irish Rooms: new words on visions for language*. Artificial Intelligence Review, 10, 1-2, 49-63.

Meehan, J. R. (1981). *TALE-SPIN*. In *Inside Computer Understanding: Five Programs Plus Miniatures*. Schank, R. C. & C. K. Riesbeck, eds. (1981), 197-226. Psychology Press.

Miller, G. A. (1991). *WordNet: An online lexical database*. International Journal of Lexicography, 3, 4.

Minerva, F. & A. Rorheim (2017). *What are the ethical consequences of immortality technology?* Aeon. https://aeon.co/ideas/what-are-the-ethical-consequences-of-immortality-technology

Minsky, M. L. (1974). *A Framework for Representing Knowledge*. MIT-AI Laboratory Memo 306. Reprinted in The Psychology of Computer Vision, ed. P. Winston (1975). McGraw-Hill.

———— (1986). *The Society of Mind*. Simon & Schuster.

———— (2006). *The Emotion Machine: Commonsense Thinking, Artificial Intelligence, and the Future of the Human Mind*. Simon & Schuster.

Mitchell, W., Wray, L. R., & Watts, M. (2016). *Modern Monetary Theory and Practice: An Introductory Text*. Amazon Digital Services LLC.

Monroe, D. (2018). *AI, explain yourself*. Communications ACM, 61, 11, 11-13.

Montague, R. (1973). *The Proper Treatment of Quantification in Ordinary English*. In: *Approaches to Natural Language*, ed. J. Hintikka, J. Moravcsik, P. Suppes, (eds.), 221-242. Dordrecht. Reprinted in Thomason (1974), 247-270.

Moravec, H. (1995). *Roger Penrose's Gravitronic Brains: A Review of Shadows of the Mind by Roger Penrose*. Psyche, 2, 6.

———— (1998). *Robot: Mere Machine to Transcendent Mind*. Oxford University Press.

Mosler, W. (2010). *The 7 Deadly Innocent Frauds of Economic Policy*. Valance Co. Inc.

Mulhall, S. (2007). *Wittgenstein's Private Language: Grammar, Nonsense and Imagination in Philosophical Investigations, §§ 243-315*. Clarendon Press Oxford.

Müller, V. C. & N. Bostrom (2016). *Future progress in artificial intelligence: A survey of expert opinion*. In *Fundamental Issues of Artificial Intelligence*, ed. V C. Müller, 555-572. Springer.

Murphy, G. L. (2004). *The Big Book of Concepts*. The MIT Press.

Navigli, R. (2009). *Word Sense Disambiguation: A Survey*. ACM Computing Surveys, 41, 2, Article 10.

Nesbitt, M. & D. Samuel (1995). *From staple crop to extinction? The archaeology and history of the hulled wheats*. S. Padulosi, K. Hammer & J. Heller (Eds.) Proceedings First International Workshop on Hulled Wheats, 41-101, July 1995, Castelvecchio Pasoli, Tuscany, Italy.

Newell, A. (1973). *You can't play 20 questions with nature and win: projective comments on the papers of this symposium*. In *Visual Information Processing* ed. W. G. Chase. Academic Press.

———— (1982). *The knowledge level*. Artificial Intelligence, 18, 87-127.

———— (1990). *Unified Theories of Cognition*. Harvard University Press.

Newell, A., J. C. Shaw, & H. A. Simon (1957). *Empirical explorations with the Logic Theory machine: A case study in heuristics*. In *Computers and Thought* ed. E. A. Feigenbaum & J. Feldman, 109-133. McGraw-Hill..

———— (1963). *Chess-playing programs and the problem of complexity*. In *Computers and Thought*, ed. E. A. Feigenbaum & J. Feldman, 39-70. McGraw-Hill.

Newell, A. & H. A. Simon (1963). *GPS, a program that simulates human thought* In *Computers and Thought*, ed. E. A. Feigenbaum & J. Feldman, 279-293 McGraw-Hill.

———— (1972). *Human Problem Solving*. Prentice-Hall.

———— (1976). *Computer science as empirical inquiry: symbols and search.* Communications ACM, 19, 3, 113-126.

Nilsson, N. J. (1983). *Artificial intelligence and the need for human labor.* Proceedings International Joint Conference on Artificial Intelligence 1983, 1195-1196.

———— (1984). *Artificial Intelligence, Employment, and Income.* The AI Magazine, 5, 2, Summer, 5-14.

———— (2005). *Human-Level Artificial Intelligence? Be Serious!* The AI Magazine, 26, 4, Winter, 68-75.

———— (2007). *The Physical Symbol System Hypothesis: Status and Prospects,* in *50 Years of Artificial Intelligence,* M. Lungarella *et al.* (eds.). Springer-Verlag, 9-17.

Nirenburg, S. & V. Raskin (2004*). Ontological Semantics.* The MIT Press.

Nirenburg, S. & Y. A. Wilks (2001). *What's in a Symbol: Ontology, Representation and Language.* Journal of Experimental and Theoretical Artificial Intelligence, 13, 1, 9-23.

Norman, D. A. (2004). *Emotional Design: Why We Love (or Hate) Everyday Things.* Basic Books.

Omohundro, S. M. (2008). *The basic AI drives.* In *Artificial General Intelligence 2008: Proceedings of the First AGI Conference,* ed. P. Wang, B. Goertzel, & S. Franklin, 483-492.

Ortony, A., G. L. Clore, & A. Collins (1988). *The Cognitive Structure of Emotions.* Cambridge University Press.

Pearl, J. (1988). *Probabilistic Reasoning in Intelligent Systems: Networks of Plausible Inference.* Morgan Kaufmann Publishers.

———— (2009). *Causality: Models, Reasoning, and Inference.* Second Edition. Cambridge University Press.

Peirce, C. S. (CP) *Collected Papers of C. S. Peirce* edited by C. Hartshorne, P. Weiss, & A. Burks, 8 vols., 1931-1958. Harvard University Press, Cambridge, MA.

Pennachin, C. & B. Goertzel (2007). *Contemporary Approaches to Artificial General Intelligence.* In *Artificial General Intelligence,* Goertzel & Pennachin, eds., 1-30.

Penrose, R. (1989). *The Emperor's New Mind: Concerning Computers, Minds and The Laws of Physics.* Oxford University Press.

———— (1994). *Shadows of the Mind: A Search for the Missing Science of Consciousness.* Oxford University Press.

———— (1996). *Beyond the Doubting of a Shadow: A Reply to Commentaries on Shadows of the Mind.* Psyche, 2, 23.

Penrose, R., A. Shimony, N. Cartwright, & S. Hawking, edited by M. Longair (1997). *The Large, the Small, and the Human Mind.* Cambridge University Press.

Penrose, R. & S. Hameroff (1995). *What 'Gaps'? Reply to Grush and Churchland.* Journal of Consciousness Studies, 2, 2, 1995, 99-112.

———— (2011). *Consciousness in the Universe: Neuroscience, Quantum Space-Time Geometry and Orch OR Theory.* In *Consciousness and the Universe,* ed. Penrose, Hameroff, & Kak, 3-42.

Penrose, R., S. Hameroff, & S. Kak, eds. (2011). *Consciousness and the Universe.* Cosmology Science Publishers, Cambridge, MA.

Pereira, F. C. (2007). *Creativity and Artificial Intelligence: A Conceptual Blending Approach.* Mouton de Gruyter, New York.

Perez y Perez, R. (1999). *MEXICA: a Computer Model of Creativity in Writing.* Ph.D. Thesis, University of Sussex.

Perez y Perez, R. & M. Sharples (2004). *Three Computer-Based Models of Storytelling: BRUTUS, MINSTREL and MEXICA.* Knowledge-Based Systems, 17, 1, 15-29.

Peterson, A. (2013). *On the internet, no one knows you're a bot. And that's a problem.* The Washington Post, August 13, 2013.

Piaget, J. (1926). *The Language and Thought of the Child.* London: Routledge & Kegan Paul.

Picard, R. (1997). *Affective Computing*. MIT Press.

Pinker, S. (1994). *The Language Instinct*. HarperCollins Publishers. Originally published by William Morrow and Co.

———— (2002). *The Blank Slate: The Modern Denial of Human Nature*. Penguin Books.

———— (2007). *The Stuff of Thought: Language as a Window into Human Nature*. Penguin Books.

———— (2014). *The Sense of Style: The Thinking Person's Guide to Writing in the 21ˢᵗ Century*. Penguin Books.

———— (2018). *Enlightenment Now: The Case for Reason, Science, Humanism, and Progress*. Penguin Books.

Pissanetzky, S. (2011). *Emergent inference, or how can a program become a self-programming AGI system?* AGI-11 Workshop on Self-Programming in AGI Systems.

Pollock, J. L. (1990). *How To Build A Person*. Bradford/MIT.

———— (1995). *Cognitive Carpentry: A Blueprint For How To Build A Person*. MIT Press.

Prawitz, D. (1965). *Natural Deduction: A Proof-Theoretical Study*. Dover Publications, 2006.

Quirk, R., S. Greenbaum, G. Leech, & J. Svartvik (1985). *A Comprehensive Grammar of the English Language*. Longman Group Limited.

Randers, J. (2012.) *2052: A Global Forecast for the Next Forty Years*. Chelsea Green Publishing.

Rapaport, W. J., S. C. Shapiro, & J. M. Wiebe (1997) *Quasi-indexicals and knowledge reports*. Cognitive Science, 21, 1, 63-107.

Redd, E. & A. S. Younger (2017). *A mathematical and physical base for 'A Standard Model of the Mind*. AAAI 2017 Fall Symposium Series Technical Reports, FSS-17-05, pp.431-436.

Reich, R. B. (2009). *Manufacturing Jobs Are Never Coming Back*. Forbes. com, May 28, 2009.

Bibliography

———— (2011). *Aftershock: The Next Economy and America's Future*. Vintage Books.

———— (2012). *Is technology to blame for chronic unemployment?* Marketplace, October 10, 2012.

Ridley, M. (1996). *The Origins of Virtue*. Viking (Penguin Books).

Rifkin, J. (1995). *The End of Work: The Decline of the Global Labor Force and the Dawn of the Post-Market Era*. G. P. Putnam's Sons, New York.

———— (2011). *The Third Industrial Revolution: How Lateral Power Is Transforming Energy, the Economy, and the World*. Palgrave MacMillan.

Rolfe, R. M. & B. A. Haugh (2017). *Integrated cognition: A framework proposal*. 2017 AAAI Fall Symposium Series Technical Reports, FS-17-05, 437-442.

———— (2017). *Integrated cognition: A survey of systems*. 2017 AAAI Fall Symposium Series Technical Reports, FS-17-05, 443-448.

Rosenbloom, P. S., A. Demski, & V. Ustun (2016). *The Sigma cognitive architecture and system: towards functionally elegant grand unification*. Journal of Artificial General Intelligence, 7, 1-103.

Rosenthal, D. M. (2005). *Consciousness and Mind*. Oxford University Press.

Russell, S. J. & P. Norvig (2010). *Artificial Intelligence: A Modern Approach*. Third Edition. Prentice-Hall.

Sacks, O. (1989). *Seeing Voices: A Journey into the World of the Deaf*. Vintage Books.

Sag, I. A., T. Wasow, & E. M. Bender (2003). *Syntactic Theory: A Formal Introduction*. Second Edition. CSLI Publications, Stanford.

Sánchez-Valencia, V. M. (1991). *Studies on Natural Logic and Categorial Grammar*. Ph.D. Thesis, University of Amsterdam.

Schank, R. C. (1973). *Identification of Conceptualizations Underlying Natural Language*. In *Computer Models of Thought and Language*, ed. Schank, R. C. & K. M. Colby, 187-247. W. H. Freeman.

—— (1999). *Dynamic Memory Revisited*. Cambridge University Press.

Schank, R. C. & K. M. Colby, eds. (1973). *Computer Models of Thought and Language*. W. H. Freeman & Co.

Schank, R. C. & R. P. Abelson (1977). *Scripts, Plans, Goals and Understanding: An Inquiry into Human Knowledge Structures*. Lawrence Erlbaum.

Schank, R. C. & C. K. Riesbeck, eds. (1981). *Inside Computer Understanding: Five Programs Plus Miniatures*. Lawrence Erlbaum.

Schank, R. C., A. Kass, & C. K. Riesbeck, eds. (1994). *Inside Case-Based Explanation*. Lawrence Erlbaum.

Scheinerman, E. R. (2013). *Invitation to Dynamical Systems*. Dover.

Schlenoff, C., J. Albus, E. Messina, A. J. Barbera, R. Madhavan, & S. Balakirsky (2006). *Using 4D/RCS to Address AI Knowledge Integration*. AI Magazine. Volume 27 Number 2, 71-81.

Schmid, H-J. (1997). *Constant and Ephemeral Hypostatization: Thing, Problem and other "Shell Nouns"*. In Proceedings 16th International Congress of Linguists, ed. B. Caron. Pergamon, CD-ROM.

—— (2000). *English Abstract Nouns as Conceptual Shells: From Corpus to Cognition*. Mouton de Gruyter.

—— (2007). *Non-compositionality and emergent meaning of lexico-grammatical chunks: A corpus study of noun phrases with sentential complements as constructions*. Zeitschrift für Anglistik und Amerikanistik 55, 3, 313-340.

Schmidhuber, J. (1987). *Evolutionary Principles in Self-Referential Learning (On Learning How to Learn: The Meta-Meta... Hook)*. Diploma thesis, Institut f. Informatik, Tech. Univ. Munich, 1987.

—— (2007). *Gödel machines: Fully Self-Referential Optimal Universal Self-Improvers*. In *Artificial General Intelligence*, ed. Goertzel & Pennachin (2007), 199-226.

Schneider, S. (2011). *The Language of Thought: A New Philosophical Direction*. MIT Press.

Schubert, L. K. (2013). *NLog-like Inference and Commonsense Reasoning*. To appear in A. Zaenen, V. de Paiva, & C. Condoravdi (eds.), *Semantics for Textual Inference*, CSLI.

Schubert, L. K., B. Van Durme, & M. Bazrafshan (2010). *Entailment Inference in a Natural Logic-like General Reasoner*. Common Sense Knowledge Symposium (CSK-10), AAAI 2010 Fall Symposium Series, Arlington, VA, November 11-13, 2010.

Searle, J. R. (1969). *Speech Acts: An Essay in the Philosophy of Language*. Cambridge University Press.

———— (1979). *Expression and Meaning: Studies in the Theory of Speech Acts*. Cambridge University Press.

———— (1980). *Minds, Brains, and Programs*. The Behavioral and Brain Sciences, 3, 417-424. Cambridge University Press.

———— (1990). *Is the Brain a Digital Computer?* Presidential address, Proceedings American Philosophical Association, 1990. Reprinted in Searle (1992).

———— (1992). *The Rediscovery of the Mind*. The MIT Press.

———— (2004). *Mind: A Brief Introduction*. Oxford University Press.

Shapiro, S. (1998). *Incompleteness, Mechanism, and Optimism*. The Bulletin of Symbolic Logic, 4, 3, 273-302.

———— (2003). *Mechanism, truth, and Penrose's new argument*. Journal of Philosophical Logic, 32, 19-42.

Shoham, Y. & K. Leyton-Brown (2008). *Multiagent Systems: Algorithmic, Game-Theoretic, and Logical Foundations*. Cambridge University Press.

Shrager, J. & P. Langley (1990). *Computational Models of Scientific Discovery and Theory Formation*. Morgan Kaufmann.

Silverman, P. & C. Whitney (2012). *Leonardo's Lost Princess: One Man's Quest to Authenticate an Unknown Portrait by Leonardo Da Vinci*. Wiley.

Singh, P. (2003). *Examining the Society of Mind*. Computing and Informatics, 22, no. 5, 521-543.

———— (2005). *EM-ONE: An Architecture for Reflective Commonsense Thinking*. Ph.D. Thesis, MIT.

Skaba, W. (2011). *Heuristic Search in Program Space for the AGINAO Cognitive Architecture*. AGI-11 Workshop on Self-Programming in AGI Systems.

Sloman, A. (1971). *Interactions between philosophy and AI: The role of intuition and non-logical reasoning in intelligence*. In Proceedings 2nd International Joint Conference on Artificial Intelligence. Reprinted in Artificial Intelligence, 2, 3-4, 209-225, and in J.M. Nicholas, ed. (1977) *Images, Perception, and Knowledge*. Dordrecht-Holland: Reidel.

———— (1978). *The Computer Revolution in Philosophy: Philosophy Science and Models of Mind*. The Harvester Press.

———— (1979). *The Primacy of Non-Communicative Language*. In *The analysis of Meaning: Informatics 5*, Proceedings ASLIB/BCS Conference, Oxford, March 1979, ed. M. MacCafferty and K. Gray.

———— (2008). *What evolved first: Languages for communicating? Or Languages for thinking?*

http://www.cs.bham.ac.uk/research/projects/cogaff/talks/glang-evo-ai1.pdf

Smith, B. C. (1982). *Procedural Reflection in Programming Languages*. Ph.D. Thesis, MIT.

———— (1996). *On the Origin of Objects*. MIT Press.

Sommers, F. (1982). *The Logic of Natural Language*. Clarendon Press.

Sowa, J. F. (1984). *Conceptual Structures: Information Processing in Mind and Machine*. Addison-Wesley, Reading, Mass.

———— (1999a). *Knowledge Representation: Logical, Philosophical and Computational Foundations*. Brooks/Cole Publishing Co., Pacific Grove, Calif.

———— (1999b). *Signs, Processes and Language Games: Foundations for Ontology*. Invited lecture, International Conference on the Challenge of Pragmatic Process Philosophy, University of Nijmegen.

———— (2001). *Semantic Foundations of Contexts.*
http://users.bestweb.net/~sowa/ontology/contexts.htm

———— (2007a). *Fads and fallacies about logic.* IEEE Intelligent Systems, March 2007, 22:2, 84-87.

———— (2007b). *Language Games, A Foundation for Semantics and Ontology.* Published in A-V. Pietarinen, ed. (2007) *Game Theory and Linguistic Meaning*, Elsevier, 17-37.

———— (2010). *The Role of Logic and Ontology in Language and Reasoning.* In *Theory and Applications of Ontology: Philosophical Perspectives*, ed. R. Poli & J. Seibt, 231-263. Berlin: Springer.

———— (2011). *Cognitive Architectures for Conceptual Structures.* Proc. 19th International Conference on Conceptual Structures, ed. S. Andrews, S. Polovina, R. Hill & B. Akhgar, LNAI 6828, 35-49. Heidelberg: Springer.

Sowa, J. F. & A. K. Majumdar (2003). *Analogical reasoning.* In *Conceptual Structures for Knowledge Creation and Communication*, Proceedings of ICCS 2003, LNAI 2746, ed. de Moor, Lex, Ganter, 16-36. Springer-Verlag, Berlin.

Sperber, D. & D. Wilson (1986). *Relevance: Communication and Cognition.* Blackwell.

Steels, L. & J. de Beule (2006). *A (very) brief introduction to Fluid Construction Grammar.* Proceedings ACL HTCL workshop on scalable language processing. New York, USA.

Steiner, G. (2011). *The Poetry of Thought: From Hellenism to Celan.* New Directions Publishing, New York.

———— (2013). *Universitas.* Lustrum Lecture, November 6, 2012. The Nexus Institute, Tilburg University.

Stewart, T. C. & C. Eliasmith (2017). *Continuous and parallel: challenges for a Standard Model of the Mind.* AAAI 2017 Fall Symposium Series Technical Reports, FSS-17-05, pp.466-469.

Sun, R. (1997). *Learning, action, and consciousness; a hybrid approach towards modeling consciousness.* Neural Networks, 10, 7, 1317-1331.

———— (2002). *Duality of the Mind: A Bottom-Up Approach to Cognition.* Lawrence Erlbaum.

Sussman, G. J. & G. L. Steele Jr. (1975). *Scheme: An Interpreter for Extended Lambda Calculus.* MIT AI Lab, AI Memo 349, December 1975.

———— (1980). *CONSTRAINTS—A Language for Expressing Almost-Hierarchical Descriptions.* Artificial Intelligence, 14, 1, 1-39.

Swartout, W., J. Gratch, R. W. Hill, E. Hovy, S. Marsella, J. Rickel, & D. Traum (2006). *Toward Virtual Humans.* AI Magazine. Volume 27 Number 2, 96-108.

Talmy, L. (2000). *Toward a Cognitive Semantics: Volume I: Concept Structuring Systems.* MIT Press.

———— (2000). *Toward a Cognitive Semantics: Volume II: Typology and Process in Concept Structuring.* MIT Press.

Tegmark, M. (2017). *Life 3.0: Being Human in the Age of Artificial Intelligence.* Alfred A. Knopf.

Thomason, R., ed. (1974). *Formal Philosophy: Selected Papers by Richard Montague.* Yale University Press.

Thórisson, K. R. (2012). *A New Constructivist AI: From Manual Methods to Self-Constructive Systems.* In *Theoretical Foundations of Artificial General Intelligence*, Wang & Goertzel, eds., 145-171.

Tomasello, M. (2003). *Constructing a Language: A Usage-Based Theory of Language Acquisition.* Harvard University Press.

Turing, A. M. (1936). *On Computable Numbers, with an Application to the Entscheidungsproblem.* Proceedings London Mathematical Society, Ser. 2, 42, 230-265.

———— (1938). *On Computable Numbers, with an Application to the Entscheidungsproblem: A Correction.* Proceedings London Mathematical Society, Ser. 2, 43, 544-546.

———— (1947). *Lecture to the London Mathematical Society on 20 February 1947.* In *The Essential Turing: Seminal Writings in Computing, Logic, Philosophy, Artificial Intelligence, and Artificial Life plus The Secrets of Enigma*, B. J. Copeland (ed.), Oxford University Press, 2004.

Bibliography

——— (1950). *Computing machinery and intelligence.* Mind, 59, 433-460.

Turner, S. R. (1994). *The Creative Process: A Computer Model of Storytelling.* Lawrence Erlbaum.

Tyler, A. & V. Evans (2003). *The Semantics of English Prepositions: Spatial Scenes, Embodied Meaning, and Cognition.* Cambridge University Press.

Ulam, S. (1958.) *Tribute to John von Neumann.* Bulletin of the American Mathematical Society, 64, 3, 1-49.

Valiant, L. G. (2013). *Probably Approximately Correct: Nature's Algorithms for Learning and Prospering in a Complex World.* Basic Books.

van Benthem, J. (1986). *Essays in Logical Semantics.* D. Reidel Publishing.

——— (1987). *Meaning: Interpretation and Inference.* Synthese 73:3, 451-470.

——— (1991). *Language in Action: Categories, Lambdas, and Dynamic Logic.* Elsevier Science B.V., Amsterdam.

——— (2008). *A Brief History of Natural Logic.* In *Logic, Navya-Nyaya & Applications, Homage to Bimal Krishna Matilal,* ed. M. Chakraborty, B. Löwe, M. N. Mitra, & S. Sarukkai. College Publications, London.

Vanderschraaf, P. & G. Sillari (2007). *Common Knowledge.* Stanford Encyclopedia of Philosophy. http://plato.stanford.edu/entries/common-knowledge/

Vinge, V. (1993). *The coming technological singularity: how to survive in the post-human era.* Whole Earth Review, Winter 1993.

Vogel, C. (1996). *Human Reasoning with Negative Defaults.* In *Practical Reasoning: Lecture Notes in Artificial Intelligence,* ed. D. Gabbay & H. J. Ohlbach, 606-621. Springer.

——— (2001). *Dynamic Semantics for Metaphor.* Metaphor and Symbol, 16, 1&2, 59-74.

——— (2010). *Group Cohesion, Cooperation and Synchrony in a Social Model of Language Evolution.* In *Development of Multimodal Interfaces: Active Listening and Synchrony,* ed. A. Esposito, N. Campbell, C. Vogel, A. Hussain, A. Nijholt, 16-32. Springer.

————— (2011). *Genericity and Metaphoricity Both Involve Sense Modulation*. In *Affective Computing and Sentiment Analysis: Emotion, Metaphor and Terminology*, ed. Khurshid Ahmad, 35-51. Springer.

Vogel, C. & J. Woods (2006). *A Platform for Simulating Language Evolution*. In *Research and Development in Intelligent Systems XXIII*, ed. M. Bramer, F. Coenen, & A. Tuson, 360-373. Springer.

Vogt, P. (2000). *Lexicon Grounding on Mobile Robots*. Ph.D. Thesis, Vrije Universiteit Brussel.

————— (2003). *Anchoring of semiotic symbols*. Robotics and Autonomous Systems, 43, 2, 109-120.

————— (2005). *The emergence of compositional structures in perceptually grounded language games*. Artificial Intelligence, 167, Issues 1-2, 206-242.

————— (2012). *Exploring the robustness of cross-situational learning under Zipfian distributions*. Cognitive Science, 36, 4, 726-739.

Vogt, P. & F. Divina (2007). *Social symbol grounding and language evolution*. Interaction Studies, 8, 1, 31-52.

Vogt, P. & E. Haasdijk (2010). *Modelling social learning of language and skills*. Artificial Life, 16, 4, 289-310.

Von Mises, L. (1949). *Human Action: A Treatise on Economics*. Yale University.

Vygotsky, L. (2012). *Thought and Language*. Edited and translated by E. Hanfmann, G. Vakar, and A. Kozulin. Foreword by A. Kozulin. MIT Press.

Walsh, T. (2017). *The singularity may never be near*. AI Magazine, 38, 3, 58-62.

Wang, H. (1996). *A Logical Journey: From Gödel to Philosophy*. The MIT Press.

Wang, P. (2011). *Behavioral Self-Programming by Reasoning*. AGI-11 Workshop on Self-Programming in AGI Systems.

Bibliography

Wang, P. & B. Goertzel, eds. (2012). *Theoretical Foundations of Artificial General Intelligence*. Atlantis Press.

Weizenbaum, J. (1966). *ELIZA: A computer program for the study of natural language communication between man and machine*. Communications ACM, 9, 1, 36-45.

———— (1984). *Computer Power and Human Reason: From Judgment to Calculation*. Penguin Books.

Weyhrauch, R. W. (1980). *Prolegomena to a Theory of Mechanized Formal Reasoning*. Artificial Intelligence, 13, 1-2, 133-170.

Whitehead, A. N. (1929). *Process and Reality*. Macmillan, New York. Corrected Edition, ed. D. R. Griffin & D. W. Sherburne, 1978, The Free Press.

Wierzbicka, A. (1996). *Semantics: Primes and Universals*. Oxford University Press.

Wilks, Y. A. (1972). *Grammar, Meaning and the Machine Analysis of Language*. Routledge.

———— (1975a). *Preference semantics*. In *The Formal Semantics of Natural Language*, ed. E. Keenan. Cambridge University Press.

———— (1975b). *An intelligent analyzer and understander of English*. Communications ACM, 18, 264-274.

———— (1975c). *A preferential pattern-seeking semantics for natural language inference*. Artificial Intelligence, 6, 53-74.

———— (2007). *In Memoriam: Karen Spärck Jones*. IEEE Intelligent Systems, May/June 2007, 22-23.

———— (2008). *On Whose Shoulders?* Computational Linguistics, 34, 4, 1-16.

———— (2017). *Will there be superintelligence and would it hate us?* AI Magazine, 38, 4, 65-70.

Wilks, Y. A., B. M. Slator, & L. M. Guthrie (1996a). *Electric Words: Dictionaries, Computers, and Meanings*. The MIT Press.

Wilks, Y. A., J. Barnden, & J. Wang (1996b). *Your metaphor or mine: Belief ascription and metaphor interpretation*. In *Discourse & Meaning: Essays in Honor of Eva Hajičová*, ed. B. H. Partee & P. Sgall, 141-161. John Benjamins Publishing Co.

Wilson, D. & D. Sperber (2012). *Meaning and Relevance*. Cambridge University Press.

Wittgenstein, L. (1922). *Tractatus Logico-Philosophicus*. Translated by C. K. Ogden, with assistance from G. E. Moore, F. P. Ramsey, and Wittgenstein. Published by Routledge & Kegan Paul. Reprinted by Dover, 1999.

——— (1953). *Philosophical Investigations*. Translated by G. E. M. Anscombe. Wiley-Blackwell Publishers, Oxford. Second Edition, 1958.

Wray, L. R. (1998). *Understanding Modern Money: The Key to Full Employment and Price Stability*. New Economic Institute.

Wright, Ian (2000). *The Society of Mind Requires an Economy of Mind*. Proceedings AISB'00 Symposium *Starting from Society: the Application of Social Analogies to Computational Systems*, Birmingham, UK: AISB, 113-124.

Yudkowsky, E. (2007). *Levels of Organization in General Intelligence*. In *Artificial General Intelligence*, Goertzel & Pennachin (2007), 389-501.

——— (2008). *Artificial intelligence as a positive and negative factor in global risk*. In *Global Catastrophic Risks*, ed. N. Bostrom & M. M. Ćircović, 308-345. Oxford University Press.

Zholkovsky, A. K. (1964). *The vocabulary of purposeful activity*. Massinyj Perevod I Prikladnaja Lingvistika, 8, 67-108.

Zhong, N. (2009). *Computational unsolvability of domain of attractions of nonlinear systems*. Proceedings of the American Mathematical Society, 137, 2773-2783.

Index

Page numbers followed by f indicate figures